Methods and Tools in Biosciences and Medicine

Microinjection,
edited by Juan Carlos Lacal, Rosario Perona and James Feramisco, 1999
DNA Profiling and DNA Fingerprinting,
edited by Jörg Epplen and Thomas Lubjuhn, 1999
Animal Toxins – Facts and Protocols,
edited by Hervé Rochat and Marie-France Martin-Eauclaire, 2000
Methods in Non-Aqueous Enzymology,
edited by Munishwar Nath Gupta, 2000
Methods for Affinity-Based Separations of Enzymes and Proteins,
edited by Munishwar Nath Gupta, 2002
Techniques in Molecular Systematics and Evolution,
edited by Rob DeSalle, Gonzalo Giribet and Ward Wheeler, 2002
Analytical Biotechnology,
edited by Thomas G.M. Schalkhammer, 2002
Prokaryotic Genomics,
edited by Michel Blot, 2003
Techniques in Prion Research,
edited by Sylvain Lehmann and Jacques Grassi, 2004

Techniques in Prion Research

Edited by
Sylvain Lehmann
Jacques Grassi

Birkhäuser Verlag
Basel · Boston · Berlin

Editors

Sylvain Lehmann
Institut de Génétique Humaine
CNRS U.P.R. 1142
141, rue de la Cardonille
F – 34396 Montpellier Cedex 5
France

Jacques Grassi
Service de Pharmacologie et d'Immunologie
Bâtiment 136, CEA/Saclay
F – 91191 Gif sur Yvette Cedex
France

Library of Congress Cataloging-in-Publication Data
Techniques in prion research / edited by Sylvain Lehmann, Jacques Grassi.
    p. ; cm. – (Methods and tools in biosciences and medicine)
  Includes bibliographical references and index.
  ISBN 3-7643-2224-1 (soft cover : alk. paper) – ISBN 3-7643-2415-5 (hard cover : alk. paper)
    1. Prions–Laboratory manuals. 2. Prion diseases–Laboratory manuals. I. Lehmann,
Sylvain, 1965– II. Grassi, Jacques, 1951– III. Series.
  [DNLM: 1. Prions–isolation & purification–Laboratory Manuals. 2. Prions–pathogenicity–
Laboratory Manuals. 3. Disease Models, Animal–Laboratory Manuals. 4. Immunohisto-
chemistry–methods–Laboratory Manuals. 5. Prion Diseases–etiology–Laboratory Manuals.
QU 25 T2548 2004]
QR502.T43 2004
579.2'9–dc22
       2004052997

Bibliographic information published by Die Deutsche Bibliothek
Die Deutsche Bibliothek lists this publication in the Deutsche Nationalbibliografie;
detailed bibliographic data is available in the Internet at <http://dnb.ddb.de>.

ISBN 3-7643-2415-5 Birkhäuser Verlag, Basel – Boston – Berlin
ISBN 3-7643-2224-1 Birkhäuser Verlag, Basel – Boston – Berlin

© 2004 Birkhäuser Verlag, P.O. Box 133, CH-4010 Basel, Switzerland
Part of Springer Science+Business Media
Printed on acid-free paper produced from chlorine-free pulp. TCF ∞
Cover illustr.: Crystal structure of the molecular complex between ovine PrP and Anti-PrP Fabs.
Printed in Germany
ISBN 3-7643-2415-5 (Hardcover)
ISBN 3-7643-2224-1 (Softcover)

9 8 7 6 5 4 3 2 1   www.birkhauser.ch

Contents

List of Contributors

ANDRÉOLETTI, OLIVIER, UMR INRA-ENVT 1225 IHAP, 23 chemin des Capelles, F – 31076 Toulouse Cedex 03, France; e-mail: o.andreoletti@envt.fr

BOYLE, AILEEN, Institute for Animal Health, Neuropathogenesis Unit, Ogston Building, West Mains Road, Edinburgh EH9 3JF, UK; e-mail: aileen.boyle@bbsrc.ac.uk

BRUCE, MOIRA E., Institute for Animal Health, Neuropathogenesis Unit, Ogston Building, West Mains Road, Edinburgh EH9 3JF, UK; e-mail: moira.bruce@bbsrc.ac.uk

CASTILLA, JOAQUIN, Protein Misfolding Disorders Laboratory, Neurology Department, University of Texas Medical Branch (UTMB), Galveston, Texas 77555, USA

CHABRY, JOELLE, Institut de Pharmacologie Moléculaire et Cellulaire, CNRS, 660, route des lucioles, F – 06560 Valbonne, France; e-mail: chabry@ipmc.cnrs.fr

CHIESA, ROBERTO, Dulbecco Telethon Institute (DTI) and Istituto di Ricerche Farmacologiche "Mario Negri", Via Eritrea 62, I – 20157 Milan, Italy; e-mail: chiesa@marionegri.it

CRÉMINON, CHRISTOPHE, Service de Pharmacologie et d'Immunologie, Bâtiment 136, CEA/Saclay, F – 91191 Gif sur Yvette Cedex, France; e-mail: christophe.cremion@cea.fr

ELFRINK, KERSTIN, Institut für Physikalische Biologie, Heinrich-Heine-Universität, D – 40225 Düsseldorf, Germany; e-mail: elfrink@biophys.uni-duesseldorf.de

FIORITI, LUANA, Dulbecco Telethon Institute (DTI) and Istituto di Ricerche Farmacologiche "Mario Negri", Via Eritrea 62, I – 20157 Milan, Italy; e-mail: fioriti@marionegri.it

FORLONI, GIANLUIGI, Istituto di Ricerche Farmacologiche "Mario Negri", Via Eritrea 62, I – 20157, Milan, Italy; e-mail: forloni@marionegri.it

GRASSI, JACQUES, Service de Pharmacologie et d'Immunologie, Bâtiment 136, CEA/Saclay, F – 91191 Gif sur Yvette Cedex, France; e-mail: Grassi@dsvidf.cea.fr

HAWKINS, STEPHEN A. C., Veterinary Laboratories Agency, Woodham Lane, New Haw, Addlestone, Surrey KT15 3NB, UK; e-mail: s.a.c.hawkins@vla.defra.gsi.gov.uk

HOPE, JAMES, VLA Lasswade, International Research Centre, Pentlands Science Park, Bush Loan, Penicuik, Midlothian, EH26 OPZ, United Kingdom; e-mail: j.hope@vla.defra.gsi.gov.uk

KIRBY, LOUISE, Institute for Animal Health, Compton Laboratories, Newbury, Berkshire RG20 7NN, United Kingdom; e-mail: louise.kirby@bbsrc.ac.uk

LEHMANN, SYLVAIN, Institut de Génétique Humaine, CNRS U.P.R. 1142, 141, rue de la Cardonille, F – 34396 Montpellier Cedex 5, France; e-mail: Sylvain.Lehmann@igh.cnrs.fr

McCONNELL, IRENE, Institute for Animal Health, Neuropathogenesis Unit, Ogston Building, West Mains Road, Edinburgh EH9 3JF, UK; e-mail: irene.mcconnell@bbsrc.ac.uk

PERRIER, VERONIQUE, Institut de Génétique Humaine, CNRS U.P.R. 1142, 141, rue de la Cardonille, F – 34396 Montpellier Cedex 5, France; e-mail: Veronique. Perrier@igh.cnrs.fr

RAYMOND, GREGORY J., National Institutes of Health, NIAID, Rocky Mountain Laboratories, Hamilton, MT 59840, USA; e-mail: graymond@niaid.nih.gov

REZAEI, HUMAN, Virologie et Immunologie Moléculaires (VIM), INRA, F-78352 Jouy-en-Josas, France; e-mail: rezaei@jouy.inra.fr

RIESNER, DETLEV, Institut für Physikalische Biologie, Heinrich-Heine-Universität, D – 40225 Düsseldorf, Germany; e-mail: riesner@biophys.uni-duesseldorf.de

SAÁ, PAULA, Protein Misfolding Disorders Laboratory, Neurology Department, University of Texas Medical Branch (UTMB), Galveston, Texas 77555, USA

SALMONA, MARIO, Istituto di Ricerche Farmacologiche "Mario Negri", Via Eritrea 62, I – 20157 Milan, Italy; e-mail: salmona@marionegri.it

SOLASSOL, JEROME, Institut de Génétique Humaine, CNRS U.P.R. 1142, 141, rue de la Cardonille, F – 34396 Montpellier Cedex 5, France; e-mail: j-solassol@chu-montpellier.fr

SOTO, CLAUDIO, Protein Misfolding Disorders Laboratory, Neurology Department, University of Texas Medical Branch (UTMB), Galveston, Texas 77555, USA; e-mail: clsoto@utmb.edu

STACK, MICHAEL J., TSE Molecular Biology Department, Veterinary Laboratories Agency, Woodham Lane, Addlestone, Surrey KT15 3NB, United Kingdom; e-mail: m.j.stack@vla.DEFRA.gsi.gov.uk

TAGLIAVINI, FABRIZIO, Istituto Neurologico "Carlo Besta", Via Celoria 11, I – 20133 Milan, Italy; e-mail: f.tagliavini@istituto-besta.it

TAYLOR, DAVID M., SEDECON 2000, Edinburgh EH13 9DX, UK; e-mail: david.taylor@sedecon2000.freeserve.co.uk

WELLS, GERALD A. H., Veterinary Laboratories Agency, Woodham Lane, New Haw, Addlestone, Surrey KT15 3NB, UK; e-mail: g.a.h.wells@vla.defra.gsi.gov.uk

Abbreviations

ACN	acetonitrile	LB	lysis buffer
AEC	3-amino-9-ethylcarbazole	LDH	lactate dehydrogenase
BCA	bicinchoninic acid	LMSB	Laemmli sample buffer
BCIP	5-bromo-4-chloro-3-indolyl phosphate	LRS	lymphoreticular system
		mAbs	monoclonal antibodies
BLB	brain lysis buffer	MALDI	matrix-assisted laser deso-rption ionization
BME	basal medium Eagle		
BSA	bovine serum albumin	MEM	minimal essential medium
BSE	bovine spongiform encephalo-pathy	MES buffer	2-Morpholino-ethanesulfonic buffer
CD	circular dichroism	MOPBS	20 mM MOPS pH 7, 150 mM NaCl
CDI	conformation dependent im-munoassay		
		MOPS	3-(N-Morpholino)propane-sul-fonic acid
CHO	chinese hamster ovary		
CJD	Creutzfeldt-Jakob disease	MS	mass spectrometry
CL3	containment level 3 (laboratory)	MTT	3-(4,5-dimethylthiazol-2-yl)-2,5-diphenyl tetrazolium bro-mide
CNS	central nervous system		
CSF	cerebrospinal fluid		
DAB	3,3'-diaminobenzidine	NaDCC	sodium dichloroisocyanurate
DCM	dichloromethane	NaPTA	sodium phosphotungstic acid
DMEM	Dulbecco modified Eagle med-ium	NBH	normal brain homogenate
		NBT	nitroblue tetrazolium
DMF	N,N-Dimethyl-formamide	NEM	N-ethyl-maleimide
DOC	desoxycholate	NMR	nuclear magnetic resonance
DPBS	Dulbecco's phosphate buffered saline	NTA	nitriloacetic acid
		OD	optical density
DPH	1,6-diphenyl-1,3,5-hexatriene	β-OGP	β-octylglucopyranosid
DTT	dithiothreitol	OIE	Office International des Epi-zooties
E.coli	Escherichia coli		
EDTA	ethylene diamine tetra acetic acid	p.i.	post inoculation
		PAGE	polyacrylamide gel electro-phoresis
EIA	enzyme immuno assay		
FBS	fetal bovine serum	PBS	phosphate buffered saline
FP	fluorescence polarization	PCR	polymerase chain reaction
GD	gravity-displacement	PI	phosphatidyl inositol
GPI	glycosyl-phosphatidyl-inositol	PK	proteinase-K
GSS	Gerstmann-Sträussler-Schein-ker	PL	porous-load
		PLP	paraformaldehyde-lysine-peri-odate
HE	haematoxylin and eosin		
HMP	hydroxymethylphenoxy (Wang-type HMP) resin	PMCA	protein misfolding cycling am-plification
i.c.	intracerebral	PMSF	phenyl-methansulfonide fluor-ide
i.p.	intraperitoneal		
IB	inclusion bodies	PNGase F	Peptide-N^4-(N-acetyl-β-glucosa-minyl)asparagin-Amidase
IMAC	immobilized metal affinity chromatography		
		PPE	personal protective equipment
IMH	infectious material homogenate	PrP	prion protein
IPTG	iso-propyl-beta-D-thio-galacto-pyranoside		

PrP106-126 synthetic peptide homologous to human PrP sequence 106–126

PrP27–30 protease cleaved PrP

PrP33–35 full length, non-protease treated PrP

PrP82–146 synthetic peptide homologous to human PrP sequence 82–146

PrP^C cellular isoform of the prion protein

PrP^{mut} mutated PrP

PrP^{res} protease-resistant isoform of the prion protein

PrP^{Sc} abnormal, disease associated isoform of PrP

PrP^{sen} protease-sensitive isoform of the prion protein

PrP^{VRQ} VRQ allele of PrP

RP-HPLC reverse-phase high-performance liquid chromatography

rPrP recombinant PrP

RT room temperature

SAFs scrapie-associated fibrils

sCJD sporadic Creutzfeldt-Jakob disease

SDS sodium dodecyl sulphate

SEAC spongiform encephalopathy advisory committee

SLB spleen lysis buffer

sPrP106 synthetic miniprion, mouse PrP Δ23–88; Δ141–176, with epitope for monoclonal antibody 3F4

SSBP1 sheep scrapie brain pool 1

TBS tris buffered saline

TFA trifluoroacetic acid

Tg transgenic

TSE transmissible spongiform encephalopathy

vCJD variant Creutzfeldt-Jakob disease

VLA Veterinary Laboratories Agency

Zw 3–12 Zwittergent 3–12

Transmissible Spongiform Encephalopathies or Prion Diseases – General Introduction

Sylvain Lehmann and Jacques Grassi

Transmissible spongiform encephalopathies (TSEs) form a group of fatal neurodegenerative disorders represented principally by Creutzfeldt-Jakob disease (CJD), Gerstmann-Sträussler-Scheinker syndrome (GSS), and fatal familial insomnia (FFI) in humans, and by scrapie and bovine spongiform encephalopathy (BSE) in animals [1]. Also called prion diseases, TSEs have the unique property of being infectious, sporadic or genetic in origin [2]. They are characterised by a silent asymptomatic period, which may be very long (up to 40 years in humans), in the absence of any specific immune or inflammatory response. During the clinical phase of the disease, specific lesions (spongiosis, astrocytosis) which constitute the veritable signature of these diseases are observed in the central nervous system. The outcome is always fatal for the affected animal or person, whose central nervous system inevitably contains, at the clinical stage of the disease, the transmissible agent which is able to infect another individual of the same species. The nature of this unconventional transmissible agent has not yet been fully elucidated. However, TSEs are almost always accompanied by the accumulation of an abnormal form of a protein in the central nervous system naturally produced by the host, the prion protein, PrP. This abnormal form (called PrPSc, for scrapie PrP) sometimes accumulates in brain as amyloid plaques or deposit of which it is the major component. Following the work of Stanley Prusiner [3], a considerable number of experimental findings were accumulated indicating that PrPSc could well be the infectious agent itself [4]. PrPSc derives from the normal PrP form (PrPC for cellular PrP) through post-translational modifications which induce a conformational change and confer on PrPSc a partial resistance to degradation by proteases, as well as a marked insolubility in the presence of detergents leading to the formation of large aggregates.

Even though there is not yet a definitive demonstration that PrPSc constitutes, by itself, the agent of TSE, there is no doubt that PrP plays a critical role in the development of TSEs and that the transition from PrPC to PrPSc is a crucial pathogenic event. The central role of PrP in TSE is also illustrated first by the strong genetic linkages between mutations in the PrP gene and genetic forms of

Methods and Tools in Biosciences and Medicine
Techniques in Prion Research, ed. by S. Lehmann and J. Grassi
© 2004 Birkhäuser Verlag Basel/Switzerland

TSE in human [1]. In addition, polymorphisms in the human and sheep PrP genes were shown to influence the transmissibility of TSEs in these species. It was also established that the efficiency of experimental transmission of the disease from one species to another is dependent on the similarity of the PrP sequences between the species [4]. This is finally exemplified by the fact that PrP-null mice are resistant to the disease [5] while transgenic mice overexpressing PrP are developing the disease at an accelerated pace. Taken together, TSEs appear as very strange transmissible diseases with uncommon properties for an infectious disease: transmissible agent incompletely identified, absence of immune response, and a very strong genetic determinism. This atypical picture is completed by the presence, in human, of sporadic forms which are not related to any known infectious event or any mutation in the PrP gene.

Until a recent past, TSEs were not considered as a critical problem in terms of public health since sporadic CJD, which is the more frequent human form, affects about one person per million inhabitants per year very uniformly at the surface of the globe [2]. Scrapie in sheep and goats has been described since 18th century and is still considered non-transmissible to humans. However, the major BSE epizootic in Great Britain (over 180,000 cases confirmed to date), and especially the announcement in 1996 of possible transmission of this disease to humans in the form of variant Creutzfeldt-Jakob disease (vCJD), generated increasing concern among European consumers. Whereas the BSE epizootic is clearly on the decline in Great Britain since 1993, it is still high in other European countries like France, Portugal, Eire, Germany, Italy and Spain. Since 1999, the testing of populations of animals at risk or of animals slaughtered normally has identified cases of BSE in various countries that considered themselves BSE-free (Germany, Spain, Italy, Denmark, Finland, Greece, Slovenia, Czech Republic, Slovakia, Austria, Japan and Canada). It is becoming more and more obvious that BSE is now a worldwide problem.

Since 1996, accumulating evidence has confirmed a link between BSE and vCJD, and a steady rise in the number of people suffering from this disease. The first epidemiological models predicted the contamination of several tens of thousands of people who may develop the full-blown disease in the coming decades [6]. The most recent forecasts are much less pessimistic, and foresee a few hundred to a few thousand victims [7, 8]. This risk of primary transmission is compounded by a risk of secondary transmission within the human species. In people with vCJD (unlike the sporadic form) it appears increasingly clear that the infectious agent is present in detectable amounts outside the central nervous system, notably in the lymph organs [9, 10], thus raising the fear of possible blood-borne transmission. This worry is heightened by the demonstration in sheep that the BSE agent may be transmitted by blood transfusion [11, 12] and the recent identification of two vCJD cases who received a blood transfusion from a donor who later died of vCJD [13, 14].

As a consequence there has been, since 1996, a significant increase in the amount of research work devoted to the field of TSEs with a need to develop or improve the techniques specifically adapted to this domain. The aim of this book is to describe in a simple and comprehensive way the more commonly used of these techniques. Taking into account the pivotal role played by PrP it is not surprising that many Chapters deal with the purification (Chapters 2 to 4), the detection (Chapters 7 to 9) and the characterization (Chapters 12 to 14) of this protein. In addition, *in vitro* (Chapters 12 and 14), cellular (Chapter 6) and animal models (Chapters 5 and 10) specifically adapted to the study of TSEs, as well as biosafety procedures (Chapter 11) are described.

References

1 Collinge J (2001) Prion diseases of humans and animals: their causes and molecular basis. *Annu Rev Neurosci* 24: 519–550

2 Parchi P, Gambetti P (1995) Human prion diseases. *Curr Opin Neurol* 8: 286–293

3 Prusiner SB (1982) Novel proteinaceous infectious particles cause scrapie. *Science* 216: 136–144

4 Prusiner SB, Scott MR, Dearmond SJ, Cohen FE (1998) Prion protein biology. *Cell* 93: 337–348

5 Bueler H, Aguzzi A, Sailer A et al. (1993) Mice devoid of PrP are resistant to scrapie. *Cell* 73: 1339–1347

6 Ghani AC, Donnelly CA, Ferguson NM, Anderson RM (2000) Assessment of the prevalence of vCJD through testing tonsils and appendices for abnormal prion protein. *Proc R Soc Lond B Biol Sci* 267: 23–29

7 d'Aignaux JN, Cousens SN, Smith PG (2001) Predictability of the UK variant Creutzfeldt-Jakob disease epidemic. *Science* 294: 1729–1731

8 Valleron AJ, Boelle PY, Will R, Cesbron JY (2001) Estimation of epidemic size and incubation time based on age characteristics of vCJD in the United Kingdom. *Science* 294: 1726–1728

9 Wadsworth JD, Joiner S, Hill AF et al. (2001) Tissue distribution of protease resistant prion protein in variant Creutzfeldt-Jakob disease using a highly sensitive immunoblotting assay. *Lancet* 358: 171–180

10 Ironside JW (2000) Pathology of variant Creutzfeldt-Jakob disease. *Arch Virol Suppl* 16: 143–151

11 Houston F, Foster JD, Chong A et al. (2000) Transmission of BSE by blood transfusion in sheep. *Lancet* 356: 999–1000

12 Hunter N, Foster J, Chong A et al. (2002) Transmission of prion diseases by blood transfusion. *J Gen Virol* 83: 2897–2905

13 Llewelyn CA, Hewitt PE, Knight RS et al. (2004) Possible transmission of variant Creutzfeldt-Jakob disease by blood transfusion. *Lancet* 363: 417–421

14 Peden AH, Head MW, Ritchie DL et al. (2004) Preclinical vCJD after blood transfusion in a PRNP codon 129 heterozygous patient. *Lancet* 364: 527–528

2 Purification of PrP^C

Kerstin Elfrink and Detlev Riesner

Contents

1 Introduction

By far most *in vitro* studies on the prion problem are carried out with recombinant prion protein (recPrP). RecPrP is available in large quantities, in high purity, from many species and with or without mutations. Although it can be used for many studies, it cannot be used for all studies. Since the cellular PrP (PrP^C) carries the post-translational modifications, i. e., the N-glycosylation at the two sites Asn 181 and Asn 197 (in hamster) and the C-terminal glycosyl-phosphatidylinositol anchor, whereas recombinant PrP is lacking those modifications, both molecules are chemically different, and a series of studies is either directed to the differences or may be affected by the differences.

For example, early studies concentrated on modifications which might be specific for the pathogenic form of PrP (PrP^{Sc}). Neither in the sequence nor in the modifications were differences found [1, 2, 3]. Other studies tried to confirm or question the structural identity of the protein moiety of recPrP and PrP^C. Only recently those studies were carried out by nuclear magnetic resonance (NMR) and did not result in significant differences. The influence of the glycosyl groups

Methods and Tools in Biosciences and Medicine
Techniques in Prion Research, ed. by S. Lehmann and J. Grassi
© 2004 Birkhäuser Verlag Basel/Switzerland

Donald Ferrier Limited

** MAIL ORDER **

ORDER NO. 011654   05/05/2005

37643   BIRKHAUSER VERLAG AG

3764324155   Lehmann, Sylvain (Ed
                        PB
                Techniques in Prion Research

CUSTOMER AC0003968002 National CJD Surveillance Unit
DATE ORDER PLACED 27/04/2005   ORDER No. PC02763

                1

NATIONAL CJD   PC 02763
OP14445

on the conversion process, i. e., the transformation of PrP^C into PrP^{Sc}, was clearly measurable but its functional relevance is unknown. So far, infectivity could not be generated *de novo* in the conversion experiments, and the reasons for that are not really known. Particularly for that reason the conversion experiments have to be performed under optimum conditions and with the optimal molecules. PrP^C represents such an optimal molecule since it is chemically identical with PrP^{Sc} but free of any infectivity, and even a very low titer of newly generated infectivity due to the conversion would be detectable. Finally the cellular function of PrP^C, i. e., the function in the non-infected host is not clear and at present is the subject of many investigations. As soon as reliable biochemistry on the whole molecule of PrP^C is carried out, it has to be done with PrP in the natural cellular form.

From the few examples given it should be evident that not only recombinant PrP but also PrP^C with its post-translational modifications has to be available in purified form. In this Chapter purification protocols are described for two sources of PrP: Firstly from hamster brain which is really the natural source, and secondly from transformed Chinese Hamster Ovary (CHO)-cells carrying several PrP genes which express acceptable amount of PrP with post-translational modifications similar to those of PrP^C from the brain.

2   Materials

Chemicals
- Cell culture media and related products: Gibco BRL-Life Technologies.
- ProteinG-Agarose: Pierce.
- Monoclonal antibody 3F4: Senetek Research.

Others from standard sources (including SIGMA-Aldrich and ICN).

Materials and equipment
- chromatography system ÄctaPrime: Amersham Pharmacia Biotech.
- IMAC column 100 × 4.6 mm, 20 μm particle diameter: PerSeptive Biosystems, Cambridge/Pharmacia Biotech.
- centriprep, centricons: Amicon Inc.
- circular dichroism: Jasco J715.

3   Methods

3.1   Method I: PrP^C from hamster brain

Several procedures have been described to purify natural PrP^C from hamster brain [1–3]. However, the best efficiency has been achieved by a semi-preparative chromatographic method [4] which will be outlined in this paragraph.

1. The first step of the protocol is homogenization of approximately 40 g hamster brain (approximately 40 animals) followed by low speed centrifugation at 3,000 xg for 10 min at 4 °C in saline buffer.
2. The pellet is rehomogenized and centrifuged again at 3,000 xg for 10 min at 4 °C. The supernatants are combined and centrifuged through a 0.85 M sucrose cushion at 100,000 xg to separate myelin. This pellet consists of a microsomal/synaptosomal membrane fraction containing PrP^C.
3. PrP^C is then solubilized by adding 2% β-octylglucopyranosid (β-OGP) and stirring at room temperature for 30 min followed by high speed centrifugation at 100,000 xg for 60 min at 15 °C. The resulting supernatant contains PrP^C which is further purified with the following chromatographic steps:

- For cation-exchange HPLC the sample is adjusted to 30 mM (0.84%) β-OGP, 150 mM NaCl (pH 7.5) and applied onto the column. The chromatography is developed with a gradient from 150 to 500 mM NaCl in 80 min. Most proteins are eluted in the flow through while PrP^C is strongly bound to the column. PrP^C-elution starts from ~300 mM NaCl. Therefore, the fraction from 300 to 500 mM NaCl is collected for further purification.
- In the next step an immobilized metal chelate affinity HPLC (IMAC) (100 × 4.6 mm, 20 μm particle diameter, PerSeptive Biosystems, Cambridge) is used. The column is charged with Co^{2+} ions according to the manufacturer's recommendations and equilibrated with 30 mM (0.84%) β-OGP, 500 mM NaCl and 0.5 mM imidazole. The sample is adjusted to 500 mM NaCl, 0.5 mM imidazole and loaded onto the column. The binding strength of PrP^C to different metal-ions follows the order: Cu^{2+} > Ni^{2+} > Zn^{2+} > Co^{2+}. Therefore, PrP^C is eluted at low imidazole concentrations with a midpoint at 7 mM imidazole.

By densitometric scanning of a silver-stained SDS-PAGE the grade of purity of the eluted fraction is estimated to be 60%. To increase the purity of PrP^C a size-exclusion HPLC is added to the protocol. Eluted fractions from IMAC are concentrated on centriprep 30 and applied using a injection loop. The column is developed with 12.5 mM MES, 12.5 mM Tris, 150 mM NaCl, 30 mM (0.84%) β-OGP, pH 7.5. PrP^C is eluted in a single peak at a position which would be expected for monomeric PrP^C. In the void volume less than 5% of PrP^C was eluted, probably as the result of nonspecific aggregation. The efficiency of this method of purification and the recovery of PrP^C are summarized in Table 1.

Troubleshooting

It is possible to use high resolution gel filtration to study protein unfolding because the retention time on SEC and the Stoke's radius of a protein are well correlated. With increasing concentration of GdnHCl to 4 M purified PrP^C undergoes a conformational change, assuming a more expanded state. The Stoke's radius shifts from 26.1 to 38.7 Å. The Stoke's radius of 26.1 Å corresponds to that expected from the molecular weight of PrP^C indicating that purified PrP^C is of a compact and native-like state and undergoes unfolding with increasing concentration of GdnHCl.

Table 1 Purification efficiency and recovery of PrP^C from hamster brain

Hamster brain	Amount of purified PrP^C	Grade of purity	Recovery
~ 40 g (~ 40 hamsters)	232.4 µg	95.4%	16.6%

3.2 Method II: PrP^C from CHO-cells

Since the required amount of animals for PrP^C–purification is rather high it was important to develop an eukaryotic expression system that renders high yield of post-translationally modified PrP^C. In 1997 a mammalian cell line of transgenic CHO cells with a glutamine synthetase selection and amplification system was established that generates PrP^C at high levels of expression [5].

Protocol 1 Cell-culture system

1. The CHO-cell-suspension culture grows at 37 °C and 2% CO_2 in a serum free modified CHO-S-SFMII-medium (Fig. 1). The following commercial solutions are added to the medium to have a 1 × end-concentration: non essential amino acids, sodium pyruvate, fungizone, insuline, transferin, selenium S supplement, adenosine (0.7 mg/ml), guanosine (0.7 mg/ml), cytidine (0.7 mg/ml), uridine (0.7 mg/ml), thymidine (0.24 mg/ml), glutamate (6 mg/ml), asparagine (6 mg/ml) and L-methionine-sulfoximine (0.09 mg/ml) and filtered (0.22 µm) directly before use.
2. Every 3–4 weeks the medium was also adjusted to 1 × penicillin/streptomycin.
3. For optimal growth CHO-cells should have a concentration of 4–8 × 10^5 cells/ ml. In the case of lower cell concentration CO_2 is upregulated to 6%.
4. For storage and further use the cells can be pelleted and stored at –20 °C.
5. For PrP^C preparation 4 × 10^9 CHO-cells (~7 g) are applied.

Figure 1 Growth curve of the CHO cell line.
The cells were diluted to a concentration of 4–8 × 10^5 cells/ml for optimal growth leading to an increase in suspension-volume.

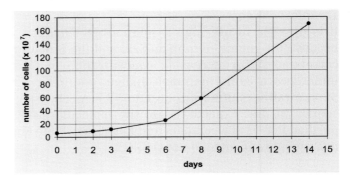

Protocol 2 PrP^C solubilization

1. To solubilize the membrane-anchored PrP^C using mild detergents, cells are stirred at 4 °C for 2 h in 50 ml solubilization buffer with 0.5% Desoxycholate (DOC) and 0.5% NP40 at neutral pH (10 mM Na-Phosphate, pH 7.2) in the presence of protease-inhibitors (1 µM Pepstatin, 1 µM Leupeptin, 2 mM PMSF, 1 mM EDTA).
2. The insoluble constituents are pelleted at 10,000 xg for 15 min whereby PrP^C remains in the supernatant. The recovery of this step is about 75% (data not shown).

The main purification takes place by two affinity-chromatographies: an immobilized metal chelate affinity chromatography (IMAC) using the intrinsic property of PrP^C to bind copper ions and an immuno-affinity chromatography employing anti-PrP antibody 3F4 (Fig. 2).

Protocol 3 Immobilized metal chelate affinity chromatography (IMAC)

The chromatographic steps are performed on a chromatography system (Äcta-Prime, Amersham Pharmacia Biotech) at 4 °C. The detailed IMAC-protocol is summarized in Table 2.
1. All solutions are filtered (0.22 µm) and degased before use.
2. For IMAC, a chelating sepharose fast flow column (Pharmacia Biotech) with a volume of 60 ml is used. Copper is loaded onto the column directly before use.
3. After equilibration with 1 mM imidazole the sample (adjusted to 1 mM imidazole) is loaded onto the column.
4. To release slightly bound protein the concentration of imidazole is increased stepwise first to 10 and then to 15 mM imidazole.
5. Elution of PrP^C takes place at very stringent conditions of 150 mM imidazole. To prevent PrP^C aggregation every solution contains 0.15% Zwittergent 3–12 (Zw 3–12).
6. The last step is regeneration of the column by removing copper with the chelator ethylene diamine tetra acetic acid (EDTA).

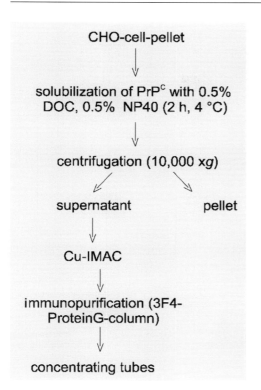

CHO-cell-pellet

solubilization of PrP^C with 0.5% DOC, 0.5% NP40 (2 h, 4 °C)

centrifugation (10,000 x*g*)

supernatant   pellet

Cu-IMAC

immunopurification (3F4-ProteinG-column)

concentrating tubes

Figure 2 Schematic purification protocol

Table 2 Detailed IMAC protocol

Solution	Volume	Flow rate	
5 mg/ml CuSO$_4$	300 ml	6 ml/min	activation
H$_2$O	200 ml	6 ml/min	washing
20 mM MOPS pH 7, 150 mM NaCl (MOPBS), 10 mM imidazole	300 ml	6 ml/min	washing
MOPBS, 1 mM imidazole, 0.15% Zwittergent 3–12 (Zw 3–12)	200 ml	4 ml/min	equilibration
sample (solubilization-supernatant), 1 mM imidazole	~50 ml	4 ml/min	loading
MOPBS, 10 mM imidazole, 0.15% Zw 3–12	until baseline	4 ml/min	washing
MOPBS, 15 mM imidazole, 0.15% Zw 3–12	until baseline	4 ml/min	washing
MOPBS, 150 mM imidazole, 0.15% Zw 3–12	until baseline	4 ml/min	elution
20 mM MOPS pH 7, 50 mM EDTA, 0.2 M NaCl, 0.15% Zw 3–12	250 ml	6 ml/min	regeneration
MOPBS, 0.02% NaN$_3$	250 ml	6 ml/min	storage

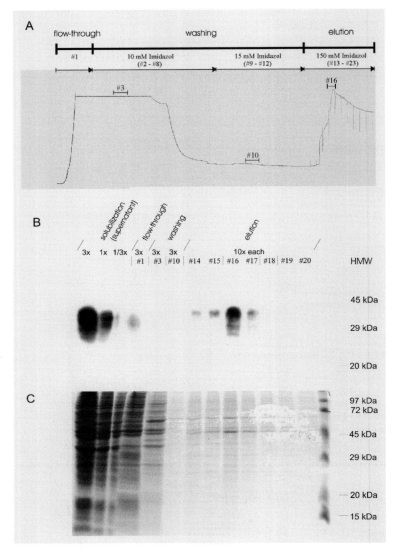

Figure 3 Analysis of a Cu-IMAC.
Supernatant of a solubilization was purified with a Cu-IMAC (for detail see Tab. 1). The chromatogram shows relative optical density (OD) units at 280 nm (A). The relative PrP^C amount in several fractions was analyzed by western-blot (3F4 as primary antibody). Relative amounts of PrP^C loaded are indicated (B). The purity was analyzed by SDS-PAGE and Coomassie-Brilliant-Blue staining (C).

Western-blot analysis showed that recovery of PrP^C in the overall eluted fractions is roughly 35% (Fig. 3 panel B). Nevertheless, Coomassie staining indicated that the grade of purity was not sufficient for further *in vitro* analysis (Fig. 3 panel C, see also Fig. 4 panel B). Therefore an immunopurification as second chromatographic step is necessary.

Protocol 4   Immunopurification

Monoclonal antibody 3F4 was provided by H. Serban and Dr. S. B. Prusiner, UCSF, San Francisco as ascites fluid. 3F4 was purified by adsorption on ProteinG-Agarose (Pierce) according to the manufacturer's recommendations. Purified 3F4 was then crosslinked with ProteinG-Agarose (Immunopure ProteinG IgG Plus Orientation Kit, Pierce). The resulting ProteinG-3F4-column has a volume of 4 ml. The detailed protocol for immunopurification is shown in Table 3.

1. To obtain high binding capacities it is recommended to decrease the flow-rate to 0.5 ml/min while the sample is loaded onto the column.
2. After removing unbound protein elution takes place by decreasing pH to 2.8 to abolish antibody antigen interaction. In contrast to the other solutions used during both chromatographies the elution buffer does not contain Zw 3–12. Since PrP^C is mainly soluble under acidic conditions it is possible to remove detergent during the acidic elution.
3. The resulting sample is diluted 4 times in Na-Acetate pH 4 and loaded into centricons with a pore size of 10 kDa (Amicon Inc.) to concentrate the protein and change the buffer to 1 mM Na-Acetate pH 4. Purified PrP^C is then fractionated and stored at –70 °C until use.

Table 3   Detailed immunopurification protocol

Solution	Volume	Flow rate	
MOPBS	100 ml	2 ml/min	washing
MOPBS, 0.15% Zw 3–12	100 ml	2 ml/min	equilibration
sample (IMAC-elution)	~100 ml	0.5 ml/min	loading
MOPBS, 0.15% Zw 3–12	until baseline	2 ml/min	washing
0.1 M HAc, 150 mM NaCl, pH 2.8	until baseline	2 ml/min	elution
1.5 M Tris/HCl, pH 8.8	until neutral pH	2 ml/min	neutralisation
MOPBS, 0.02% NaN$_3$	100 ml	2 ml/min	storage

An exemplary chromatogram of an immunopurification (Fig. 4A) shows that the elution could be divided into two elution peaks. SDS-PAGE and silver staining reveals that the second peak is of a higher purity (Fig. 4B). Therefore fractions of the second peak were concentrated in centricons. The reason for the seperation of two peaks during elution is not clear. It could be speculated from silver-stained SDS-PAGE (Fig. 4B) that different proteins interact with PrP^C corresponding to the different peaks. In comparison to diglycosylated PrP^C from hamster brain which has a molecular weight of 33–35 kDa, CHO-PrP^C becomes hyperglycosylated [5] and therefore migrates in a band at 36–39 kDa (Fig. 4C).

Table 4 shows the average efficiency of CHO PrP^C purification. Although the total amount of prepared CHO-PrP^C is rather low, the yield and grade of purity is high enough to allow spectroscopic analysis.

Table 4 Purification efficiency and recovery of PrP^C from CHO-cells

CHO-cells	Amount of purified CHO-PrP^C	Concentration of CHO-PrP^C	Purity	Recovery
4 × 10⁹ cells (7 g)	30–50 μg	0.5–1 mg/ml	96–98%	10–15%

Figure 4 Analysis of an immunopurification. The antibody 3F4 was crosslinked to a ProteinG-agarose column. This column was used for immunopurification of the previously IMAC-eluted PrP^C (for detail see Tab. 2). The chromatogram shows relative OD units at 280 nm (**A**). The grade of purity was analyzed by SDS-PAGE and silver staining (**B**). By western-blot (3F4 as primary antibody) the PrP^C corresponding bands were identified (**C**).

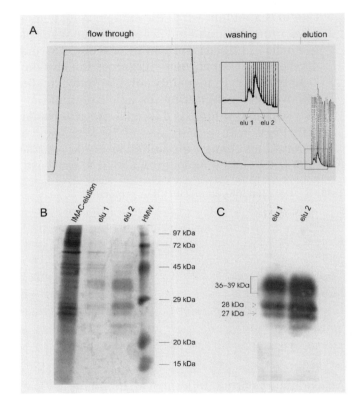

Protocol 5 Characterization of purified CHO-PrP^C

1. The secondary structure of PrP^C purified from CHO-cells was characterized by circular dichroism (Fig. 5).

   Furthermore solubility was analyzed by differential centrifugation at 100,000 xg for 1 h (Beckmann, TLA-45 rotor) (Fig. 5). The solubility assay showed that CHO-PrP^C is mainly soluble at pH 4 in the absence of detergent. Addition of low concentrations of SDS (0.05%) led to complete solubility. The secondary structure of soluble CHO-PrP^C at 0.05% SDS is of a mainly α-helical character. In the absence of SDS CHO-PrP^C has an increased content in β-structure.

   It is known that very low concentrations of detergent can induce β-structure and aggregation of PrP^C (data not shown). The increased content in β-structure and partial insolubility of the purified CHO-PrP^C can therefore be explained by incomplete removal of detergent during elution from the ProteinG-3F4-column.

2. To analyze the glycosylation pattern of CHO-PrP^C samples were digested with Peptide-N^4-(N-acetyl-β-glucosaminyl)asparagin-Amidase (PNGase F, Sigma Aldrich). This enzyme specifically cleaves N-gycosylations.

   CHO-PrP^C was adjusted to 2% SDS, boiled for 5 min, PNGase F was added at a ratio of 250 mU/ μg PrP^C and incubated at 37 °C for various times. The results are shown in Figure 6. Partial cleavage led to a shift of the main band from ~38 kDa (Fig. 6: 0 min) to three bands at 29, 24 and 23 kDa (Fig. 6: 2–5 min). After complete cleavage of the N-glycosylations two forms of CHO-PrP^C with a molecular weight of 24 and 23 kDa remained (Fig. 6: 30 min, 24 h). These two bands could correspond to unglycosylated CHO-PrP^C with and without the Glycosyl-phosphatidyl-inositol (GPI) anchor. The absence of the GPI anchor leads to a decreased SDS-PAGE mobility of about 1 kDa [6]. Therefore the band at 23 kDa should correspond to unglycosylated CHO-PrP^C containing the GPI anchor whereas the band at 24 kDa is unglycosylated CHO-PrP^C lacking the GPI anchor.

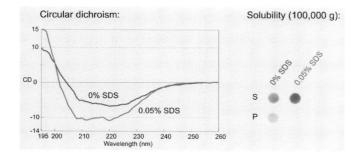

Figure 5 Characterization of purified CHO-PrP^C. PrP^C was incubated at a concentration of 1 mg/ml for 1 day at 25 °C. The secondary structure was analyzed by circular dichroism (Jasco J715). To determine the solubility samples were centrifuged at 100,000 xg for 1 h and analyzed by dot blot (3F4 as primary antibody).

Figure 6 Analysis of the glycosylation pattern of CHO-PrP^C.
Samples were adjusted to 2% SDS and boiled for 5 min. PNGase F was added at a ratio of 250 mU/μg PrP^C. After different incubation time samples were analyzed by western blot (3F4 as primary antibody).

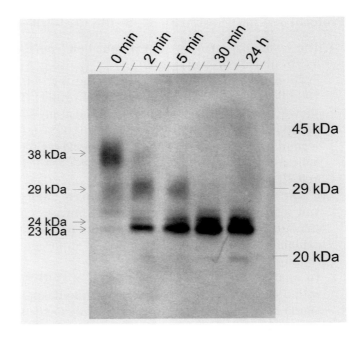

Troubleshooting

PrP^C is a membrane anchored protein with high hydrophobicity due to its C-terminal GPI-anchor. This insolubility is a problem in purifying PrP^C and it is important to prevent PrP^C aggregation. On this account every solution used during preparation contains detergent. A critical step of the described protocol is the last concentrating step because it is possible that besides PrP^C the detergent is also concentrated. This would lead to an unknown detergent concentration in the resulting sample and partial denaturing of the protein (measured by circular dichroism (CD), data not shown). Therefore it is important to remove detergent during elution of the last chromatography (immuno-purification). To make sure that residual detergent is not concentrated on the centricons the samples are diluted 4 times prior to concentration on centricons.

Another critical step is the immunopurification via the ProteinG-3F4-column. The antibody is exposed to acidic pH during elution of PrP^C. To prevent hydrolysis of the antibody the column is neutralized immediately after use. Nevertheless the column can only be used for about five preparations and efficiency depends on the number of chromatographies being performed. This also leads to the fact that the division into two elution peaks (Fig. 4) is not always obtained to the extent reported here.

References

1   Turk E, Teplow DB, Hood LE, Prusiner SB (1988) Purification and properties of the cellular and scrapie prion proteins. *Eur J Biochem* 176: 21–30

2   Bendheim PE, Potempska A, Kascsak RJ, Bolton DC (1988) Purification and partial characterization of the normal cellular homologue of the scrapie agent protein. *J Infect Dis* 158: 1198–1208

3   Pan K-M, Stahl N, Prusiner SB (1992) Purification and properties of the cellular prion protein from Syrian hamster brain. *Protein Sci* 1: 1343–1352

4   Pergami P, Jaffe H, Safar J (1996) Semi-preparative chromatographic method to purify the normal cellular isoform of the prion protein in nondenatured form. *Anal Biochem* 236: 63–73

5   Blochberger TC, Copper C, Peretz D et al. (1997) Prion protein expression in chinese hamster ovary cells using a glutamine synthetase selection and amplification system. *Protein Eng* 10: 1465–1473

6   Narwa R, Harris DA (1999) Prion protein carrying pathogenic mutations are resistant to phospholipase cleavage of their glyolipid anchor. *Biochemistry* 38: 8770–8777

3 Purification of the Pathological Isoform of Prion Protein (PrPSc or PrPres) from Transmissible Spongiform Encephalopathy-affected Brain Tissue

Gregory J. Raymond and Joëlle Chabry

Contents

1 Introduction

Main histopathological hallmarks of TSEs are severe spongiosis, gliosis and brain tissue deposition of an abnormal isoform (PrPSc or PrPres) of the normal cellular prion protein (PrPC or PrPsen). Despite identical primary sequences, PrPC and PrPSc isoforms differ in their tertiary structures [1] and as a consequence, they exhibit distinct physicochemical properties. For instance, PrPSc forms macromolecular aggregates that are insoluble in mild detergent whereas PrPC

Methods and Tools in Biosciences and Medicine
Techniques in Prion Research, ed. by S. Lehmann and J. Grassi
© 2004 Birkhäuser Verlag Basel/Switzerland

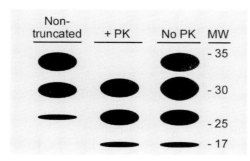

Figure 1 Diagrammatic representation of an immunoblot of PK-treated (+PK lane) and non-PK treated (no PK lane) PrPSc molecules in the 263K TSE strain preparations. The various bands represent the different glycosylation forms of PrPSc molecules. The non-truncated lane represents the PrPSc forms that would be observed in SDS-PAGE silver stain and immunoblots if no *in situ* and/ or preparation proteolysis occurred. Non-PK preparations normally contain a mixture of truncated and non-truncated forms as described in the concluding remarks, whereas PK treated preparations contain only truncated forms. Approximate molecular weights (MW) are shown on the right and expressed in kilodaltons.

is a monomeric soluble protein mostly anchored in lipid bilayer by a glycosylphosphatidyl inositol moiety. Moreover, while PrPC is sensitive to proteolysis, PrPSc aggregates are highly resistant to treatment by various proteases such as proteinase K (PK) enzyme. PK treatment results in a complete digestion of PrPC and in an amino-terminal truncation of PrPSc in which the first 90 residues are removed (referred to as PrP27–30) (Fig. 1). The purification procedure initially described by Bolton et al. [2] has been based on these fairly unique physicochemical properties of the PrPSc isoform. The PrPSc preparation procedure described here is a modified version of the Bolton et al. protocol.

Using experimental procedure described here, PrPSc preparations of several species from transmissible spongiform encephalopathy (TSE)-affected brain tissues have been successfully purified such as human (Creutzfeldt-Jakob disease (CJD) and variant-CJD (vCJD)), ovine, bovine and cervids. Purified PrPSc preparations have also been obtained from several mouse- and hamster-adapted scrapie strains such as ME7, Chandler, 87 V, 22L, 263K, hyper, and drowsy.

Numerous experiments have used PrPSc isolated by this purification procedure allowing detailed analyses of several different aspects of TSE diseases. For instance, infrared spectra experiments have demonstrated that β-sheet structure content and so spatial conformations differ when two distinct hamster-adapted scrapie strains were compared [3]. In 1994, B. Caughey and colleagues developed a cell-free conversion system demonstrating for the first time that PrPSc aggregates are capable of converting PrPsen molecules into a conformation indistinguishable from PrPres [4, 5] (see Chapter 12 in this book). Notably, information bore at least in part by PrPSc aggregates such as strain specificity and species barrier are kept during the purification procedure and so transmissible to newly converted PrP [6–10]. However, it is important to note that though PrPSc preparations purified according to the current protocol remained infectious, they were not purified to homogeneity.

2   Materials

Chemicals
- Dithiothreitol (DTT, Promega, Madison,WI, USA)
- N-tetradecyl-N,N-dimethyl-3-ammonio-1-propanesulfonate, sulfobetaine, (SB 3–14, Calbiochem or Sigma)
- N-laurylsarcosine (Sigma), protease inhibitors were from Roche
- RNAase A (from bovine pancreas, protease-free) and DNAase I were from Calbiochem or Roche
- Protein concentration assay kit (BCA) from Pierce.

Equipment
General comments: to avoid cross contamination between different scrapie strains, we recommend the use of new glassware and instruments or single use materials for animal inoculations, dissection of brains from affected animals, and during the purification of PrP^Sc. This is especially important if the isolates are to be used in animal bioassays. Avoid blood contamination during dissection of brains, rinse excised brains in PBS, flash freeze in liquid nitrogen and store at –70 °C. All solutions must be filtered (0.2 µm) and placed in lint-free vessels previously rinsed with distilled or deionized water. All procedures are performed in a laminar flow biosafety cabinet in an appropriate laboratory according to governmental regulations. Use personnel protection equipment (lab coat, gloves and face mask) and change gloves frequently.

Solutions, reagents and buffers

- 10X Phosphate-buffered saline (PBS), pH 6.9

  26.8 g $Na_2HPO_4.7H_2O$
  13.8 g NaH_2PO_4- H_2O
  75.9 g NaCl

  1000 ml

- 1X PBS, pH 7.4

  100 ml 10X PBS, pH 6.9
  900 ml deionized water

  1000 ml

- 1 M $NaPO_4$, pH 6.9

  340.5 ml 1 M Na_2HPO_4
  159.5 ml 1 M NaH_2PO_4

  500 ml

- 5X TEND
  50 mM Tris-HCl, pH 8.0 at 4 °C
  5 mM EDTA
  665 mM NaCl
  1 mM DTT (add immediately before using)

  10 ml 1 M Tris-Cl, pH 8.0
  2 ml 0.5 M EDTA
  26.2 ml 5 M NaCl
  0.002 ml 1 M DTT

  500 ml

- 10% Sarcosine (TEND)   50 g Na N-lauroyl-sarcosine   500 ml
                                          100 ml 5X TEND

- 10% NaCl, 1% SB 3–14 (TEND)   171 ml 5 M NaCl   500 ml
                                          100 ml 5X TEND
                                          5 g SB 3–14

- TMS   1 ml 1 M Tris-Cl   100 ml
  10 mM Tris-HCl, pH 7 at 20°C   0.5 ml 1 M MgCl$_2$
  5 mM MgCl$_2$ 2 ml 5 M NaCl
  100 mM NaCl

- 1 M Sucrose   68.5 g sucrose   200 ml
  100 mM NaCl,   4 ml 5 M NaCl
  0.5%SB 3–14   1 g SB 3–14
  10 mM Tris-CI, pH 7.4 at 20°C   2 ml 1 M Tris-CI

- 0.5% SB 3–14/1X PBS   0.5 g SB 3–14   100 ml
                                          10 ml 10X PBS

- 1 M DTT (1000X)

- Protease inhibitors PIs (1000X)   0.5 mg/ml Leupeptin = 1 µM
                                          (in H$_2$O)
                                          1 mg/ml Aprotinin = 0.15 µM
                                          (in H$_2$O)
                                          1 mg/ml Pepstatin = 1 µM
                                          (in methanol)
                                          24 mg/ml Pefabloc = 0.1 M
                                          (in 0.05 M Tris-HCl, pH 6.5
                                          at 20°C)

DTT and PI solutions are freshly diluted into buffers.

3   Methods

3.1   General procedure

Day 1

- Prior to beginning purification, brains were dissected from TSE-affected animals (refer to general comments, above). Brains can be stored at –70 °C for long periods (years) without apparent affect.
- Use 35 g of brain tissue. Make sure that the 10% homogenate (w/v) will fit into the rotor used for centrifugation. This procedure may be scaled down without apparent detriment to final product.
- Place the brains in a disposable Petri dish, determine mass, and then chop into small pieces using a scalpel while they are thawing.
- Transfer to a disposable beaker and rinse several times in 100 ml of cold PBS (until there is no color) to remove residual blood.
- Pour off phosphate buffered saline (PBS) and suspend to yield a 10% w/v brain suspension in cold 10% sarcosine (TEND) that has been supplemented with 1 mM DTT and protease inhibitors (DTT/PIs) just prior to use. Transfer aliquots of the suspension to a 40 ml dounce and homogenize with the loose pestle ten times and then ten times with the tight pestle. Avoid forming foam. Collect homogenated aliquots in a container on ice. Keep dounce and brain homogenate on ice as much as possible during processing. Upon completion of douncing the entire suspension, incubate on ice for 1 h. Use caution with glass dounces and pestles; check them for foreign objects and cracks, rinse them well with filtered H$_2$O, and protect eyes and hands.
- Using a 50 ml syringe with canula, transfer the brain homogenate into Beckman OPTISEAL tubes. Centrifuge for 30 min at 4 °C in a Beckman Type 50.2 Ti rotor at 15,000 rpm (22,000 xg).
- Using the canula and 50 ml syringe, transfer supernatants (S1a) to new OPTISEAL tubes avoiding pellets (P1) and ~3–4 ml of the supernatant above the pellet. Keep S1a on ice. Combine and re-extract the P1 pellets using the residual supernatant and divide equally to two OPTISEAL tubes. Fill these two tubes with cold 10% sarcosine (TEND + PIs), mix well by inversion, incubate 30 min on ice and centrifuge as above.
- Transfer supernatant (S1b) from re-extracted pellets to new OPTISEAL tubes avoiding pellets. Top off S1a and S1b tubes using cold 10% sarcosine (TEND + PIs) and centrifuge for 2 h 30 min at 4 °C at 38,000 rpm (150,000 xg) in the Type 50.2 Ti rotor.
- Using canula, remove and discard the supernatants (S2) appropriately and gently rinse the pellets (P2) two times with 5 ml room temp. 10% NaCl, 1% SB 3–14 (TEND + PIs). Resuspend and combine all the pellets and bring to a final volume of 1–2 ml per gram of starting brain material (e. g. 30–60 ml). Homogenize by douncing ten times at room temperature using a tight pestle.

- Transfer the homogenate to OPTISEAL tubes with the canula and centrifuge at 45,000 rpm (225,000 xg) for 2 h at 20 °C in the Type 50.2 Ti rotor.
- Using canula, remove and discard the supernatants appropriately and gently rinse the pellets (P3) twice with 5 ml of TMS containing PIs and then resuspend the pellets 10 ml of TMS (+PIs). Homogenize by douncing 10 times on ice with the tight pestle – use a 7 ml dounce.
- Transfer the homogenate into a 15 ml disposable polypropylene tube and incubate at 37 °C for 2 h with 100 µg/ml RNAase A. Transfer tube to 4 °C overnight.

Day 2

- Add 1 M $CaCl_2$ to 5 mM, mix, then add DNAase I to 20 µg/ml and mix again. Incubate for 2 h at 37 °C. Ca^{+2} is required for DNAase activity, however if $CaCl_2$ is added before RNAase A treatment a large amount of white precipitant will occur which is rRNA, thus the stepwise process that we recommend here.
- At this point in the preparation, a portion of the material may be digested with proteinase K (PK) enzyme. This step is optional – digestion with PK does yield purer PrP^{Sc}, however it is all truncated at the amino-terminus. Typically, we will PK digest half of the total preparation. Add PK to 10 µg/ml after the DNAase incubation, mix well, and incubate 1h at 37 °C. At the end of the incubation, add Pefabloc to 1 mM, mix, incubate on ice for 15 min, and then proceed to the next step.
- Add ethylene diamine tetra acetic acid (EDTA) to 20 mM and mix, next add NaCl to 10% w/v and mix, and then add SB 3–14 to 1%. It is important to add SB 3–14 <u>after</u> EDTA and NaCl solutions in order to avoid precipitation of free divalent cations.
- Bring the volume to 20 ml by adding TMS (+PIs) buffer. Using a syringe with canula, the suspension is layered over 10 ml of 1 M sucrose, 100 mM NaCl, 0.5% SB 3–14 and 10 mM Tris-Cl, pH 7.4, into a OPTISEAL tube (50.2) then pelleted through the sucrose cushion at 45,000 rpm (225,000 xg) for 2 h at 20 °C in the Type 50.2 Ti rotor.
- Carefully remove the supernatant, then cut away the top of the OPTISEAL tube using a scalpel. Rinse the resulting golden pellet (P4) twice with 2 ml 0.5% SB 3–14/1X PBS and resuspend in 1 ml of 0.5% SB 3–14/1X PBS, transfer the suspension to a TL100.3 polycarbonate centrifuge tube and sonicate for 1 min in ice-cold water in a cuphorn at maximum power. Centrifuge at 75,000 rpm (200,000 xg) for 15 min at 4 °C in a Beckman TL100.3 ultracentrifuge.
- Carefully remove and discard the supernatant, the pellet is finally resuspended in 1 ml of 0.5% SB 3–14/1X PBS, transferred to a microtube and sonicated as above. The protein concentration is assayed as described below and adjusted to ~1–2 mg/ml. Make and store aliquots at 4 °C for long periods of time. We have observed that storage at 4 °C is more stable than storage at –20 °C, as measured by cell-free conversion activity.

3.2 Analysis of the PrP^{Sc} preparation

Protein concentration assay (BCA)

Preparation of the samples
• Mix in a microtube: 5 µl of PrP^{Sc} preparation
                                                    2.5 µl SDS 20%
                                                    42.5 µl of distilled water
The mix (dilution 1/10) is sonicated for 1 min and boiled for 5 min.
• Mix in a microtube: 10 µl of the dilution 1/10
                                                    5 µl SDS 20%
                                                    85 µl of distilled water
The mix (dilution 1/100) is sonicated for 1 min and boiled for 5 min.

Preparation of the standard curve
• Mix in a microtube: 12.5 µl of the bovine serum al-
                                                    bumin, (BSA; 2 mg/ml)
                                                    5 µl SDS 20%
                                                    82.5 µl of distilled water
                    [BSA]f = 250 µg/ml into SDS 1%.

Tube No.	Volume of SDS 1% (µl)	Volume of BSA (250 µg/ml) (µl)
0	10	0
1	8	2
2	6	4
3	4	6
4	2	8
5	0	10

Tube No.	Volume of SDS 1% (µl)	PrP^{Sc} dilution	Volume of PrP^{Sc} preparation (µl)
6	0	1/10	10
7	6.7	1/10	3.3
8	9	1/10	1
9	0	1/100	10
10	6.7	1/100	3.3
11	9	1/100	1

- Add 200 µl of solution made of 6.25 ml of solution A and 0.125 ml of solution B according to the manufacturer's recommendations. Vortex and incubate at 65 °C for 30 min.
- Read the absorbance at 562 nm.
- The total protein concentration is usually between 2 and 10 mg/ml.
- The PrP^{Sc} is typically 60–90% of the total protein, as discussed below. To determine the purity of the PrP^{Sc} relative to contaminating proteins, SDS-PAGE silver stain analysis is necessary. Immunoblot analysis using quantified PrP standards is used to determine PrP^{Sc} yield.

SDS-PAGE analysis – Silver staining and Western blotting

Buffers
- Sample buffer (1X SABU) 62.5 mM Tris-HCl, pH 6.8 at 20 °C
  5% ultrapure glycerol
  3 mM EDTA
  5% SDS
  4% β-mercaptoethanol or 50 mM DTT
  0.02% bromophenol blue

- PK buffer 50 mM Tris-HCl, pH 8.5 at 4 °C
  1 mM CaCl$_2$
  50% ultrapure glycerol
Store PK solutions at –20 °C in either 1 mg/ml or 10 mg/ml concentrations.

- 10X TN 500 mM Tris-Cl pH 8.5 at 20 °C
  1.5 M NaCl

Preparation of the samples
Typically, the PrP^{Sc} yield from the hamster 263K strain is 50–100 µg per gram of brain. Assuming the low end of the expected yield, the preparation is diluted to 1–2 mg/ml in 0.5%SB/1X PBS. For PAGE silver stain and immunoblot analyses, an aliquot of the non-PK preparation is PK treated. Then this PK-treated and a non-PK-treated sample are diluted such that 100 ng, 30 ng, and 10 ng of the expected amount are loaded onto separate lanes of SDS-PAGE gels (Fig. 2). Preparations that have been purified using the optional PK step should be prepared without PK digestion as described below.

5 µg (estimated) = 5 µl or 2.5 µl of suspended 1–2 mg/ml PrP^{Sc} preparation (always vortex immediately before removing aliquots), as described above.
1 µl 1 mg/ml PK in PK buffer.
5 µl 10X TN and H$_2$O to 50 µl, yielding an estimated PrP^{Sc} concentration of 100 ng/ µl.
- Incubate at 37 °C for 60 min, then add 1 µl of 0.1 M Pefabloc, vortex to mix and place on ice to stop the PK digestion.

- Make an identical sample without PK (" no PK ") and keep on ice during the incubation period.
- For each sample (+PK and " no PK ") remove 10 µl of sample and add 90 of 1X sample buffer (1X SABU).This sample is estimated to be 100 ng/10 µl. Mix well by vortexing and dilute by taking 30 µl of the sample and adding 70 µl of the 1X SABU – this sample is 30 ng/10 µl. Vortex the 100 ng/10 µl sample well and dilute 10 µl into 90 µl of 1X SABU – this is 10 ng/10 µl.
- Additional samples of 100 ng, 30 ng and 10 ng of previously purified and characterized PrP^{Sc} samples or recombinant PrP^C are used as standards for quantification.
- Vortex all samples, then boil for 5 min, centrifuge in microfuge to 10,000 rpm very briefly, repeat vortexing, recentrifuge, and load 10 µl of each sample onto gel.
- Any samples remaining may be frozen at –20 °C for later gel analysis. When removing from the freezer, it is critical to immediately boil the frozen samples for 5 min. The PK has the ability to refold and become active, whereas the PrP^{Sc} does not refold and becomes susceptible to complete degradation by the PK.
- Samples were loaded on two different electrophoresis gels (10% NuPage, bis-tris, Invitrogen) and run using MES running buffer according to the manufacturer's recommendations.
- At the end of the electrophoresis, gels were either silver stained [11] or immunoblotted and then PrP bands revealed with specific PrP antibodies (Fig. 2).

4   Remarks and conclusions

The method described yields PrP^{Sc} preparations that are between 60% and 90% pure compared to total protein content. Non-PK digested preparations are usually less pure than PK-digested preparations as expected since nearly all other proteins are digested by PK. However, a significant protein contaminant that has been identified in these preparations is ferritin that survives even PK treatment (unpublished observation, Raymond GJ, Hughson AG and Caughey B). Variations in purity may depend on the TSE strain, the stage of the disease, the manner of excision and storage of the brains, and the expediency of the purification. In addition to contamination by other proteins, the final PrP^{Sc} sample most likely contains traces of lipids, nucleic acids, carbohydrates and ions (copper, iron, calcium, etc.). Non-PK treated preparations are characterized by the presence of truncated and non-truncated PrP^{Sc} forms (refer to diagram). This is likely due to *in situ* proteolysis that occurs in infected brains; however additional proteolysis may occur during brain excision and/or purification procedures, even though steps such as rapid freezing of brains, careful

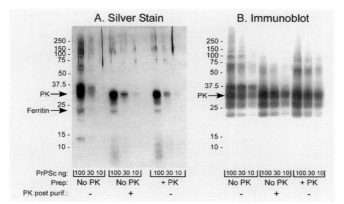

Figure 2   PrP^{Sc} purified from brain tissue of Syrian golden hamsters clinically affected with the hamster-adapted scrapie strain, 263K.
Panel A is a silver-stained SDS-PAGE gel showing the purity of PrP^{Sc} isolated using the protocol described. The migration of molecular weight markers, in kilodaltons (kd), is shown to the left of the panel. PrP^{Sc} bands are the prominent bands located between ~25–37.5 kd. Amounts loaded in nanograms (ng) are shown at the bottom of each lane. "Prep" refers to whether or not PK was used during the purification of the PrP^{Sc} and the " + " and " – " refer to PK treatment (+) or no PK treatment (-) after the purification. The top arrow at ~30 kd indicates the migration position of PK, which can be seen as the sharp band in the no PK preparation samples that were treated with PK after the purification was completed. The PK band comigrates with the doubly glycosylated PrP^{Sc} band in this gel system. Note that the sharp PK band is not present in the + PK preparation, because it was removed during the purification process. The lower arrow at ~23 kd shows the position of ferritin which is a contaminant identified in these preparations Some of the lower molecular weight bands have been identified as various PrP peptides. The non-PK digested no PK preparation sample amino terminal 90 amino acid residues are removed when PK digested, explaining the shift downward of the PrP^{Sc} bands. ***Panel B*** is an immunoblot of an SDS PAGE gel of the same samples used for the silver stain gel in Panel A developed using a 1:20,000 dilution of 3F4 ascites, a hamster-specific monoclonal antibody [12]. Though the immunoblot is more sensitive at detection of the various PrP^{Sc} molecules than the silver stain, it does not detect contaminating proteins due to its specificity for PrP. In the immunoblot, higher molecular weight immunoreactive species are detected that are likely multimers of PrP. The migration of PK is indicated by the arrow. In this gel system, since PK comigrates with the doubly-glycosylated PrP^{Sc} it does cause blockage of the antibody binding, making the band appear as a doublet. If a different gel system is used, such as a 16% tris-glycine (Invitrogen), the PK band does not comigrate with the doubly glycosylated PrP^{Sc} molecule. Note the truncation of the PK-treated PrP^{Sc} compared to the non-PK treated PrP^{Sc} preparation.

temperature control and the use of protease inhibitors during the purification are used in an attempt to prevent proteolysis. Biochemical and biophysical analyses of these preparations continues. Efforts continue to improve upon the purification of PrP^{Sc}. Since PrP^{Sc} itself is thought to be the infectious agent of TSEs, it would be of interest to achieve its purification to homogeneity in order to test this hypothesis.

References

1   Stahl N, Baldwin MA, Teplow DB et al. (1993) Structural studies of the scrapie prion protein using mass spectrometry and amino acid sequencing. *Biochemistry* 32: 1991–2002

2   Bolton DC, Bendheim PE, Marmorstein AD, Potempska A (1987) Isolation and structural studies of the intact scrapie agent protein. *Arch Biochem Biophys* 258: 579–590

3   Caughey B, Raymond GJ, Bessen RA (1998) Strain-dependent differences in beta-sheet conformations of abnormal prion protein. *J Biol Chem* 273: 32230–32235

4   Kocisko DA, Priola SA, Raymond GJ et al. (1995) Species specificity in the cell-free conversion of prion protein to protease-resistant forms: A model for the scrapie species barrier. *Proc Natl Acad Sci USA* 92: 3923–3927

5   Caughey B, Raymond GJ, Kocisko D et al. (1997) Scrapie infectivity correlates with converting activity, protease resistance, and aggregation of scrapie-associated prion protein in guanidine denaturation studies. *J Virol* 71: 4107–4110

6   Kocisko DA, Priola SA, Raymond GJ et al. (1995) Species specificity in the cell-free conversion of prion protein to protease-resistant forms: a model for the scrapie species barrier. *Proc Natl Acad Sci USA* 92: 3923–3927

7   Bessen RA, Kocisko DA, Raymond GJ et al. (1995) Non-genetic propagation of strain-specific properties of scrapie prion protein. *Nature* 375: 698–700

8   Bossers A, Belt PBGM, Raymond GJ et al. (1997) Scrapie susceptibility-linked polymorphisms modulate the *in vitro* conversion of sheep prion protein to protease-resistant forms. *Proc Natl Acad Sci USA* 94: 4931–4936

9   Raymond GJ, Hope J, Kocisko DA et al. (1997) Molecular assessment of the potential transmissibilities of BSE and scrapie to humans. *Nature* 388: 285–288

10 Raymond GJ, Bossers A, Raymond LD et al. (2000) Evidence of a molecular barrier limiting susceptibility of humans, cattle and sheep to chronic wasting disease. *Embo J* 19: 4425–4430

11 Blum H, Beier H, Gross HJ (1987) Improved silver staining of plant proteins, RNA and DNA in polyacrylamide gels. *Electrophoresis* 8: 93–99

12 Kascsak RJ, Rubenstein R, Merz PA et al. (1987) Mouse polyclonal and monoclonal antibody to scrapie-associated fibril proteins. *J Virol* 61: 3688–3693

Rapid Procedure for Purification and Renaturation of Recombinant PrP Protein

Human Rezaei

Contents

1 Introduction

PrP^C purification from cells and tissues requires high quantities of biological material, owing to the generally low cellular expression level of the protein [1, 2]. Expression in *E. coli* circumvents this difficulty. It was used as a source of full-length prion protein as well as its (90–231) C-terminal fragment for biophysical and functional studies [3–9].

Attempts to purify the full-length recombinant protein involve either expression in the periplasm of *E. coli* followed by sequential chromatographic separations [9–12], or fusion to a histidine tag and purification through binding to Ni or

Methods and Tools in Biosciences and Medicine
Techniques in Prion Research, ed. by S. Lehmann and J. Grassi
© 2004 Birkhäuser Verlag Basel/Switzerland

Cu affinity columns [8, 9, 13, 14]. The first procedure is rather tedious and the soluble protein is prone to noncontrolled proteolysis, while, in the second procedure, the histidine tag needs to be cleaved by mild digestion with thrombin [12], which might also lead to noncontrolled proteolytic cleavage of the protein itself.

Here we present a direct one-step process to purify the full-length protein expressed in a bacterial system without addition of a histidine tag (Fig. 1). This procedure, which uses the intrinsic capacity of the N-terminal octapeptides to bind divalent cations, allows the specific sorting of the full-length protein. Renaturation on the column avoids intermolecular S–S bond formation and leads to a very pure and stable protein [13]. This procedure, first set-up for the ovine prion protein, could be applied to PrP from other species, provided they possess octapeptide repeats. In addition it allows to purify the PrP N-terminal domain for the purpose of investigating its metal cation binding properties.

Figure 1 Purification procedure of the full-length recombinant PrP protein. The inclusion bodies were dissolved in buffer 1 containing 6 M urea and then loaded on Ni-Sepharose column. The washing step with urea-free buffer led to renaturation of the protein while bound to the column. The purified protein was then eluted by 1 M imidazole.

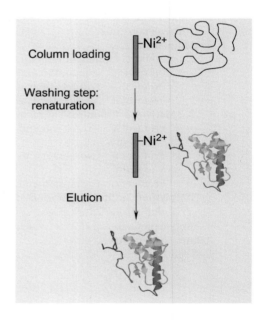

2   Materials

Chemicals
Composition of different buffer used for purification was:
- B1: 6 M urea, 500 mM NaCl, 20 mM Tris, 2.5 mM imidazole, pH 7.5
- B2: 500 mM NaCl, 20 mM Tris, 2.5 mM imidazole, pH 7.5
- B3: 500 mM NaCl, 20 mM Tris, 15 mM imidazole, pH 7.5
- B4: 6 M Guanidium chloride, 100 mM NaCl, 20 mM Tris, 2.5 mM imidazole, pH 7.5
- B5: 100 mM NaCl, 20 mM Tris, 2.5 mM imidazole, pH 7.5
- B6: 50 mM NaCl, 20 mM Tris, 15 mM imidazole, pH 7.5

Equipment
Protein purification and desalting: All purification and desalting were made using Akta FPLC chromatography equipment (Pharmacia). The purification was made with a homemade column loaded by chelating sepharose fast flow (Amersham). A G25 HiPrep26/10 column (Pharmacia) was used for the desalting procedure.

The circular dichroism (CD) spectra were recorded on thermostated Jasco 810 spectropolarimeter using a 0.5 mm-path length cuvette (Hellma). Each spectrum was an average of six scans.

3   Methods

3.1   Method 1: Cloning of the different PrP forms

Protocol 1   Human PrP

1. To obtain the cDNA of human PrP, perform a PCR amplification of human genomic DNA, using CGGAATTCCATATGAGCAAGAAGCGCCCGAAGCCTG and CGCGGATCCGTC ACGATCC TCTCTGGTAATAGGC as primers.
2. Digest the PCR product using *NdeI* and *BamHI* and clone in the corresponding restriction sites of the pET22b+ vector (Novagen) so as to generate the HuPrP plasmid. Verify the clone by sequencing and check for the presence of a methionine at codon 129 and a histidine at codon 187.

Protocol 2   Full-length ovine PrP

1. To obtain the cDNA of the $V^{136}R^{154}Q^{171}$ variant of the ovine PrP (PrP^{VRQ}), perform a PCR amplification of corresponding genomic DNA, using CGGAATTCCATATGAGCAAGAAGCGCCCGAAGCCTG and CGCGGATCCGT-CACGATCCTCTCTGGTAATAGGC as primers.
2. Digest the PCR product using *NdeI* and *BamHI* and clone in the correspond-ing restriction sites of the pET22b+ vector (Novagen) so as to generate the ovine PrP (PrP^{VRQ}) plasmid. Verify the clone by sequencing.

Protocol 3   Amino terminus of ovine PrP

1. To obtain cDNA coding for the N-terminal (23–124) part of the ovine protein, use as template, the ovine PrP (PrP^{VRQ}) plasmid of Protocol 2. Perform a PCR amplification using the following primers: 5′-AGAATTCATATGAG-CAAGCGTCCAAAACCTGGCGGAGG-3′ and 5′-ACTCGAGGATCCTATCA-CACTGCTCCAGCTGCAGCAGCT-3′.
2. Digest the PCR product using *NdeI* and *BamHI* and clone in the correspond-ing restriction sites of the pET22b+ vector (Novagen) so as to generate the ovine p23–124 PrP plasmid.Verify the clone by sequencing.

Note: For amplification, all constructs were transfected by thermal shock into the DH5α *E.coli* strain. For production, the different constructs were trans-fected by thermal shock into the BL21 DE3 *E.coli* strain (containing the polymerase T7 gene).

3.2   Method 2: Production of recombinant PrP proteins

The method described below, can be used for all mammalian PrPs having the octa-repeat sequences.

Protocol 4   Expression and purification of the full-length PrP protein

This protocol is applicable to all mammalian PrPs having the octa-repeat sequences.

1. Grow BL21 DE3 bacteria containing the various constructs in Luria Bertani medium (LB) containing 100 μg/ml ampicilin up to an optical density (OD) of 0.8 at 600 nm. Then induce the expression during 16 h with 1 mM isopropyl β-D-thio-galactopyranoside (IPTG).
2. After induction, centrifuge bacteria and resuspend its in Tris 10 mM ethylene diamine tetra acetic acid (EDTA) 10 mM buffer and lyse in the same buffer containing lysozyme (1.5 mg/ml) and Triton X100 (0.1%) during 15 min at 37 °C. This step can lead to the formation of a viscous solution.

3. Sonicate the bacteria lysate for 1 min at room temperature without pre-cooling of the solution. Centrifuge at 10,000 g for 15 min leads to the formation of a white pellet that constitutes the inclusion bodies (IB) which contain PrP protein.

4. Dissolve the IB at 4 °C in buffer B1 containing 6 M urea, 500 mM NaCl, 20 mM Tris, 2.5 mM imidazole, pH 7.5 and load it onto Ni-sepharose column (chelating sepharose fast flow, Pharmacia) at a ratio of 4 ml resin for 1 litre of culture.

5. Wash the column successively with
   – 10 ml buffer B1
   – 25 ml buffer B2
   – 25 ml buffer B3

6. Elute the PrP protein in one step, with a 10 mM 3-(N-Morpholino)propane-sulfonic acid (MOPS) pH 7.2 buffer, containing 1 M imidazole.

7. Replace the elution buffer with the desired buffer by passing through a desalting G25 column. For longterm conservation, the protein can be kept concentrated in pH 6.0 morpholinoethan-sulfonic acid (MES) buffer (10 mM). The concentration of the protein solution is calculated by measuring its OD at 280 nm using 58718 M^{-1} cm^{-1} as molar extinction coefficient.

Protocol 5   Expression and purification of the N-terminal domain of PrP

1. For the expression of the N-terminal fragment of PrP repeat the step 1 of Protocol 4.

2. After induction, centrifuge bacteria and resuspend it in Tris 10 mM EDTA 30 mM buffer and supplement with protease inhibitor cocktail (Boehringer Mannheim) to prevent degradation during bacteria lyses. This step can lead to the formation of a viscous solution.

3. Cool down the bacteria lysate at 4 °C for 10 min then sonicate for 30 s – not more – at room temperature. Centrifugation at 15,000 g for 15 min leads to the formation of a white pellet that constitutes the IB which contain the N-terminal fragment of PrP.

4. Dissolve the IB at 4 °C in buffer B1 containing 6 M Guanidium chloride, 100 mM NaCl, 20 mM Tris, 2.5 mM imidazole, pH 7.5 and load it onto a Ni-sepharose column (chelating sepharose fast flow, Pharmacia) at a ratio of 4 ml resin for 1 litre of culture.

5. Wash the column successively with
   – 10 ml buffer B4
   – 25 ml buffer B5
   – 25 ml buffer B6

6. For elution, use a 10 mM MOPS buffer pH 7.2, containing 1 M imidazole.

7. The elution buffer is then replaced by the desired buffer by passing through a desalting G25 column. The concentration of the protein solution is calculated by measuring its OD at 280 nm using 38,500 M^{-1} cm^{-1} as molar extinction coefficient.

Protocol 6   Expression and purification of soluble full-length PrP

The key step for the production of soluble PrP is the IPTG induction step. The formation of the PrP IB appears to increase with temperature. To avoid the formation of IB, the induction step was performed at 32 °C.

1. Grow BL21 DE3 bacteria containing the various constructs in Luria Bertani medium (LB) containing 100 µg/ml ampicilin up to an OD of 1.0 at 600 nm at 37 °C. Then induce the expression during 20 h with 1 mM iso-propyl-β-D-thio-galactopyranaside (IPTG) at 32 °C.
2. After induction, centrifuge bacteria and resuspend them in Tris 10 mM buffer and supplement with EDTA free protease inhibitor cocktail (Boehringer Mannheim) to prevent degradation during bacteria lyses. This step can lead to the formation of a viscous solution.
3. Cool down the bacteria lysate at 4 °C for 10 min then sonicate for 30 s – no more – at room temperature. Centrifugation at 15,000 g for 15 min leads to the formation of a brown pellet.
4. Load the supernatant onto a Ni-sepharose column (chelating sepharose fast flow, Pharmacia) at a ratio of 3 ml resin for 1 litre of culture.
5. Wash the column successively with
   – 25 ml buffer B2
   – 25 ml buffer B3
6. For elution, use a 10 mM MOPS buffer pH 7.2, containing 1 M imidazole.
7. The elution buffer was then replaced by the desired buffer by passing through a desalting G25 column). The concentration of the protein solution is calculated by measuring its OD at 280 nm using 58718 M^{-1} cm^{-1} as molar extinction coefficient.

3.3   Method 3: Characterisation of the purified PrP and the N-terminal domain

To characterise the purified PrP or related fragment, an SDS PAGE should be performed first to estimate the degree of purification or degradation that may have occurred during the purification procedure (Figs 2 and 3). Regarding the full-length PrP, a unique band must be observed at 23 kDa after elution. As shown in Figure 2C and 3B, the mass spectroscopy analysis of the eluted PrP for VRQ sheep variant and human PrP reveal the expected molecular weight.

For the N-terminal domain, some degradation was observed during the biosynthesis in *E.coli* system. After an SDS PAGE and Coomasi blue staining, at least three major bands could be obsered (Fig. 4A). The mass spectroscopy analysis of the elution product reveals multiple cleavages in the C terminal part of the N-terminal domain (Fig. 4C). However no degradation in the octa-repeat domain was observed by this purification procedure.

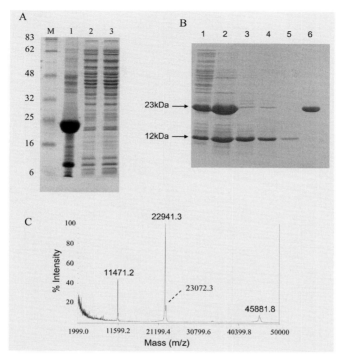

Figure 2   Characterisation by SDS-PAGE of the different stages of the purification protocol
(Coomasie blue staining).

A: Different bacterial fractions after centrifugation: lane 1, pellet containing inclusion bodies
and highly concentrated in PrP; lane 2, supernatant devoid of PrP; lane 3, total extract from
noninduced BL21 DE3 bacteria as control. Molecular weigth markers are on the left lane.

B: Different stage of purification: lane 1, whole extract of BL21 DE3 bacteria after IPTG
induction; lane 2, purified inclusion bodies (same as panel A, lane 1); lane 3, column flow
through; lanes 4 and 5, washings with buffer 2; lane 6, elution by 1 M imidazole.

C: Verification of the integrity of the purified protein by MALDI. Double- (11471.2 Da) and
mono-charged (22941.3 Da) peaks are indicated. The small peak at 23072.3 Da corresponds
to the protein where M23 has not been removed.

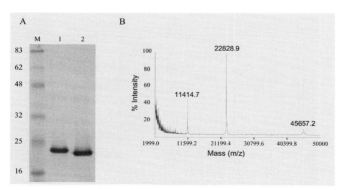

Figure 3   Purification and characterisation of HuPrP. The full-length recombinant PrP was
purified according to the same protocol as described for sheep VRQ variant.

A: SDS PAGE of 1 µg of HuPrP after purification (lane 1) and 1 µg of sheep PrP^{VRQ} (lane 2)
(Coomasie blue staining).

B: Matrix-assisted laser desorption ionisation (MALDI) spectra of the purified HuPrP showing
the expected molecular weight (22828.9)

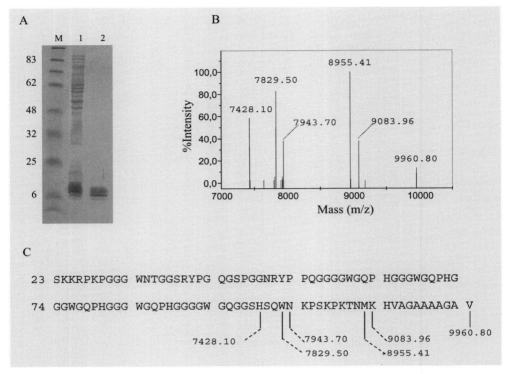

Figure 4 Characterisation of the PrP (23–124) N-terminal fragment after production in bacteria or *in vitro* transcription/translation.
A: Lane 1, SDS-PAGE of BL21 DE3 bacterial extracts after IPTG induction. Lane 2, elution of the Ni-NTA column (Coomasie blue staining). Molecular weight standards are on the left.
B: MALDI analysis of the purified (23–124) fragment. Six cleavage fragments were observed, corresponding to the sites indicated on the sequence in C.

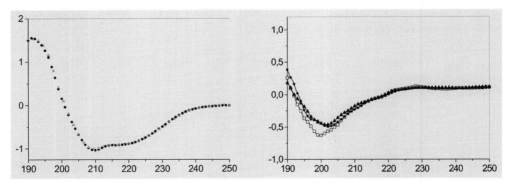

Figure 5 Secondary structure of the full-length protein produced by two different ways and comparison with the (23–124) fragment.
A: Comparison of the CD spectra of PrP purified by renaturation from inclusion bodies (open triangle) and PrP produced in soluble form and purified (filled circles).
B: CD spectra of the (23–124) PrP fragment at different pHs: open circles, pH 4.5; filled circles, pH 6.0 and filled triangles, pH 7.2.

The secondary structure of purified full-length PrP from IBs or from soluble form can be investigated by circular dichroism. As it was shown in Figure 5A, the two spectra are completely overlapping, revealing the same secondary structure. Furthermore, the spectra reveal a high content of alfa helix characteristics of PrP^C. Furthermore, the CD spectra of N-terminal domain at different pH's reveal a random coil organisation as it was previously reported [7].

4   Applications

Several types of investigation in the prion area requires large quantities of pure PrP protein. Previous purification protocols made use of 1) fusion of His-tag to the PrP sequence [12] which must then be enzymatically cleaved, or 2) tedious chromatography steps on ion exchange matrices and inverse phase [10]. These previous methods were long and laborious and usually lead to a partial degradation of the full-length PrP.

The technique presented here has many advantages. First, the expression in IBs insures a high initial concentration of the protein and its easy separation from the other cellular proteins (Fig. 2) as well as a high protection against proteases. Second, the requirement of Ni binding through octapeptide domains performs a selection of the full-length molecules and an automatic elimination of the degradation products. The imidazole concentration (1 M) necessary to compete the Ni chelation by the full-length prion protein is higher than when eluting a histidine tagged protein (20 mM to 400 mM imidazole gradient). This would lead to an efficient purification from other bacterial contaminants that might bind loosely to the Ni-sepharose column. Third, the heterogeneous phase renaturation prevents precipitation of the protein and risks intermolecular S–S bond formation. Indeed we have shown that intramolecular bonds are correctly formed [13] and that the purified protein is monodisperse and monomeric. Finally, the PrP purified by heterogeneous phase renaturation and the soluble PrP present similar secondary structures.

Last, the protocol presented here can be performed within a maximum of four days for an amount of purified full-length PrP between 20–30 mg per litre of bacteria culture.

Acknowledgements

This work was supported by funds from the French Institut National de la Recherche Agronomique (INRA).

References

1   Prusiner SB, Scott MR, DeArmond SJ et al. (1998) Prion protein biology. *Cell* 93: 337–348

2   Pergami P, Jaffe H, Safar J (1996) Semi-preparative chromatographic method to purify the normal cellular isoform of the prion protein in nondenatured form. *Anal Biochem* 236: 63–73

3   Pan KM, Baldwin M, Nguyen J et al. (1993) Conversion of alpha-helices into beta-sheets features in the formation of the scrapie prion proteins. *Proc Natl Acad Sci USA* 90: 10962–10966

4   Donne DG, Viles JH, Groth D et al. (1997) Structure of the recombinant full-length hamster prion protein PrP(29–231): the N terminus is highly flexible. *Proc Natl Acad Sci USA* 94: 13452–13457

5   Lopez Garcia F, Zahn R, Riek R, Wuthrich K (2000) NMR structure of the bovine prion protein. *Proc Natl Acad Sci USA* 97: 8334–8339

6   Billeter M, Riek R, Wider G et al. (1997) Prion protein NMR structure and species barrier for prion diseases. *Proc Natl Acad Sci USA* 94: 7281–7285

7   Li R, Liu T, Wong BS et al. (2000) Identification of an epitope in the C terminus of normal prion protein whose expression is modulated by binding events in the N terminus. *J Mol Biol* 301: 567–573

8   Brown DR, Qin K, Herms JW et al. (1997) The cellular prion protein binds copper *in vivo*. *Nature* 390: 684–687

9   Stockel J, Safar J, Wallace AC et al. (1998) Prion protein selectively binds copper(II) ions. *Biochemistry* 37: 7185–7193

10 Mehlhorn I, Groth D, Stockel J et al. (1996) High-level expression and characterization of a purified 142-residue polypeptide of the prion protein. *Biochemistry* 35: 5528–5537

11 Hornemann S, Korth C, Oesch B et al. (1997) Recombinant full-length murine prion protein, mPrP(23–231): purification and spectroscopic characterization. *FEBS Lett* 413: 277–281

12 Zahn R, von Schroetter C, Wuthrich K (1997) Human prion proteins expressed in *Escherichia coli* and purified by high-affinity column refolding. *FEBS Lett* 417: 400–404

13 Rezaei H, Marc D, Choiset Y et al. (2000) High yield purification and physico-chemical properties of full-length recombinant allelic variants of sheep prion protein linked to scrapie susceptibility. *Eur J Biochem* 267: 2833–2839

14 Viles JH, Cohen FE, Prusiner SB et al. (1999) Copper binding to the prion protein: structural implications of four identical cooperative binding sites. *Proc Natl Acad Sci USA* 96: 2042–2047

5 Animal Models of Transmissible Spongiform Encephalopathies: Experimental Infection, Observation and Tissue Collection

Gerald A. H. Wells and Stephen A. C. Hawkins

Contents

Methods and Tools in Biosciences and Medicine. Techniques in Prion Research, ed. by S. Lehmann and J. Grassi.
Birkhäuser Verlag Basel/Switzerland. © Crown copyright 2004. Published with the permission of the Controller
of Her Britannic Majesty's Stationary Office. The views expressed are those of the author and do not necessarily
reflect those of Her Britannic Majesty's Stationary Office or the VLA or any other government department.

1   Introduction

From the early nineteenth century animal models increasingly contributed to the growing understanding of the nature and causes of infectious diseases [1]. Initially animal experimentation focussed on the demonstration of transmissibility, but the concept of animal models later broadened to encompass any condition found in an animal that is of value in studying a biological phenomenon. Animal models range, therefore, from the selection for study of naturally occurring disorders in conveniently managed animal species, to the experimental reproduction of diseases in conventional, or genetically manipulated, host species, or different species. With increasing knowledge has come the universal realisation that experimental infections often differ in important respects from the natural disease they appear to mimic.

No group of disorders has in recent times presented a greater array of unexplained biological phenomena than the transmissible spongiform encephalopathies (TSEs) or prion diseases. Key historical deployment of animal experimentation in the study of the TSEs has included the initial transmission of scrapie of sheep, the archetypal transmissible spongiform encephalopathy (TSE), to sheep [2]; the recognition of similarities between kuru and scrapie [3], which was critically instrumental in initiating studies of the transmissibility of kuru; the first transmission of sheep scrapie to mice [4] and the successful transmission of kuru [5] and sporadic Creutzfeldt-Jakob disease (sCJD) [6] to apes. Emergence of an understanding of host susceptibility to transmission came from the identification of a gene [7] later shown to be the gene encoding prion protein (PrP), controlling incubation period of scrapie in mice. Development of an incubation time assay for high titre material in hamsters was important in accelerating biochemical study of the scrapie agent [8, 9]. Seminal pioneering studies of the tissue distribution of infectivity in sheep and goats with scrapie were conducted by mouse bioassay [10–12]. Bovine spongiform encephalopathy (BSE), recognised in Britain in 1986 [13], presented as a nationwide food-borne epidemic in domestic cattle [14, 15] and provided the first model of a naturally occurring TSE in which the oral route is implicit in transmission. The range of species in which natural or experimental TSE infections have been recorded and indeed, species apparently resistant to such infections (see Tab. 1 in [16]), provides an indication of the conventional wild-type animal model range that has potential in this field. Most recently an expanding array of transgenic (Tg) mouse models are providing information and many further questions about the biology of PrP.

It is not within the scope of this Chapter to outline all of the methods utilised in the study of TSEs through animal models. The aim here is rather to present, in more detail than is given in publications of experimental results, the most widely utilised basic methods, that have been employed principally at the Veterinary Laboratories Agency (VLA), United Kingdom, for the study of BSE and scrapie in mice and certain domestic animal species, notably the host species of BSE, cattle. Furthermore, here we concentrate on those procedures that are especially related to routine studies of TSE agent transmission; for the procedures concerning laboratory rodent animal house operations, general protocols for use in containment laboratories, details of anaesthesia, euthanasia and necropsy dissection procedures, readers are referred to standard texts.

2   Methods

2.1   Safe working

The safety procedures recommended here are those applied in a laboratory dealing predominantly with the BSE agent. Experimental handling of scrapie agents theoretically does not require the same level of containment and personal protection, but since in many laboratories studies are also directed toward the identification of the possible occurrence of BSE in sheep, the same safety procedures are adopted for all TSE agents. All the procedures are in accordance with National guidelines and legislation [17] and local risk assessments.

The personal protective equipment (PPE) listed here is that which has been progressively implemented over one and a half decades of working with the BSE agent; it may in some instances go beyond legal requirements, but provides a level of protection with which staff are content given the potential for exposure when dealing continuously with BSE infective material. PPE are listed below according to the main environments in which they are required:

1. Containment Level 3 laboratory (CL3 laboratory)
   Disposable parturition gown, mobcap and gloves (latex) and dedicated footwear or disposable overshoes.
2. Dosing of large animal species
   Wellington boots, disposable boiler suit, neoprene overtrousers and coat, rubber gloves sealed with tape to sleeves of the coat and airflow safety helmet.

3. Husbandry of exposed large animal species
   During the 28 day post dosing period PPE should be as worn during dosing.
   After this post dosing period: Wellington boots, neoprene overtrousers,
   disposable boiler suit and disposable gloves.

After dosing or conducting inoculations of large animals, clothing is cleaned and
decontaminated. Initially, any material from between the cleats of boots is
washed off and the exterior of all PPE is scrubbed thoroughly with sodium
hypochlorite containing 20,000 ppm available chlorine (hereafter referred to as
sodium hypochlorite). Gloves and disposable boiler suit are removed and
discarded. Discarded disposable clothing, sharps containers and syringes, etc.
are sent for immediate incineration.

Emergency responses to potential TSE contamination of operators or the
laboratory are dealt with according to national safety guidelines [17]. Cutting or
penetrating injuries which have breached the skin should be encouraged to
bleed freely and should be physically cleansed using warm, soapy water. Where
wounds are deep or where the risk of contamination is high, medical advice
must be sought in addition to the first aid measures.

Any spillages of potentially contaminated TSE material in the laboratory
must be cleaned up immediately by staff wearing appropriate PPE. The spillage
should be soaked up with absorbent material that can then be secured within
polythene bags and incinerated. Contaminated surfaces, instruments and
apparatus should be swabbed thoroughly with sodium hypochlorite or 1 molar
sodium hydroxide or autoclaved at a TSE decontamination approved temperature and pressure as appropriate. A contact time with sodium hypochlorite or
sodium hydroxide of 1 h is recommended by repeated wetting of the surface
with the disinfectant.

For further details of cleaning and decontamination of protective clothing,
post-mortem room, equipment and instruments, readers are referred to national guidelines [17].

2.2   Preparation and storage of inocula

Inocula are produced from a wide variety of tissues collected from a range of
animal species for use in the inoculation of both laboratory animal and
domestic animal species. The sources of inocula may be natural or experimental cases of BSE, scrapie or other TSE, or uninfected control animals. Fresh
tissue collected post-mortem is stored at –80 °C until required. Material from a
single tissue may be required to be pooled from several animals of a single
experiment or of a known disease phenotype. Inocula are most frequently
produced for the inoculation of mice to determine the presence of infectivity
(qualitative assay), endpoint titration or incubation period assay (quantitative
assays) or for biological strain typing studies. The procedures involved in the

last approach are dealt with in Chapter 10 and are not discussed further here. All inocula are prepared in a microbiological safety cabinet within a CL3 laboratory. When preparing inocula, aseptic technique is ensured to prevent bacteriological contamination. It is also imperative that there is no opportunity for cross-contamination of tissues; therefore tissues must be processed individually.

Protocol 1   Brain and brainstem homogenate

Brain homogenate is prepared from the whole brain of TSE infected or uninfected control animals. A homogenate of brainstem (providing slightly higher titre material than whole brain in the bovine with BSE) [18–20] is comprised of tissue inclusive of the medulla oblongata, through pons and mesencephalon. This material may be used directly for inoculation of laboratory or domestic animal species or archived to provide material for later use or for other research establishments.

1. Containers of brain material are allowed to thaw at room temperature in a CL3 laboratory and placed into the safety cabinet.
2. For preparation of small quantities (e.g. single brainstems) of homogenate, equal amounts of brain material from each case are weighed out into a sterile Petri dish. Individual pieces of brain material are forced through a sterile metal sieve with a pore size of 1.5 mm, using a sterile metal spatula. The homogenate is passed into an appropriate sterile container and then mixed well using a sterile spatula. Aliquots of the homogenate are placed into appropriate sterile containers labelled with the homogenate code, the date prepared and the weight of the homogenate.
3. For preparation of large quantities (e.g. multiple/pooled brains) of homogenate brain material is weighed to ensure equal amounts from multiple sources and either placed into a Waring laboratory variable speed 1 litre blender, or into a sterile container for homogenisation using an electric hand blender, without added fluids. The material is blended until judged sufficiently homogenised. Where large quantities of inoculum for feeding exposures are required, for example in the experimental oral exposure of pigs to the BSE agent [21], individual homogenates are mixed in large sterile vessels and aliquots repeatedly remixed by stirring until a puree of all brains is achieved. The homogenised brain material is then decanted into appropriate sterile containers and labelled as previously.
4. All containers of brain homogenate are placed in a –80 °C freezer.
5. All recyclable equipment is subject to autoclaving prior to cleaning. All waste is placed in a sealed autoclave bag for decontamination prior to incineration. Hand blenders cannot be cleaned sufficiently to be used again and should be soaked in sodium hypochlorite for one hour prior to being sent for incineration.

Protocol 2   Standard 10^{-1} dilution (and serial dilutions)

Preparing the following items for use prior to tissue processing helps to maintain sterility and ensures minimal disruption of the subsequent procedures:

- A previously sterilised gauze filter is placed into the base of the barrel of a 50 ml hypodermic syringe using sterile forceps, the plunger is replaced and the syringe is then returned to its packaging.
- The side and top of a sterile bijoux are labelled using a permanent freezer pen. Each tissue has a unique number.
- It is ensured that all consumable items required are available before starting the procedure (i. e., Petri dishes) and that recyclable items (i. e., Griffiths tubes) are available and sterile.

1. Thawed tissue is placed into a sterile Petri dish and trimmed to remove fascia, etc. before weighing. Depending upon tissue availability, organ, etc. 0.5–2 g of tissue is weighed into the Petri dish discarding any remaining tissue. (*Note: on occasions, the final tissue weight is comprised of a pool of identical tissues, in equal amounts, from several animals.*)
2. By using two sterile disposable scalpels, the tissue is macerated thoroughly.
3. The appropriate amount of sterile normal saline (0.85% NaCl), previously autoclaved at 136 °C for 30 min, is measured into a universal container using a sterile pipette to obtain a 10^{-1} dilution (w/v).
4. The macerated tissue is placed into a Griffiths tube. A minimal amount of the measured saline is added and the tissue carefully ground. More saline is added if necessary, taking care not to exceed three-quarters tube capacity, until the tissue is thoroughly homogenised. Some tissues are difficult to homogenise using this method, e. g., some peripheral nerve tissue, and should be ground until it is evident that no further homogenisation is possible.
5. The tissue suspension is decanted into a universal and topped up with the remaining saline. This is done by washing out the Griffiths tube with a small volume of the saline and pouring the contents carefully into the universal.
6. Taking previously prepared 50 ml or 20 ml syringes (each with gauze filter), the plunger is removed and the barrel placed into a sterile universal. The prepared suspension of inoculum is mixed thoroughly by shaking and poured carefully into the syringe barrel. The plunger is then replaced and depressed gently until all the liquid phase has passed through the filter into the universal. Finally, the plunger of the syringe is depressed firmly to remove as much liquid as possible remaining in the gauze filter. With some tissues (e. g., peripheral nerve) it may be necessary to remove the remaining solid tissue plug with forceps and discard, as this often blocks the filter.
7. Aliquots of the filtered suspension are measured into pre-labelled bijous using an Eppendorf pipette with sterile 5 ml tip.
8. Samples are stored at −80 °C.

9. All disposable equipment is autoclaved at 136 °C for 18 min for sterilisation prior to incineration.
10. Similarly, all instruments, Griffiths tubes and glass pipettes are autoclaved for re-use.

For endpoint titrations, serial ten fold dilutions of 10^{-1} inocula are prepared.

Bacteriological screening of inocula

All inocula produced are screened for bacteriological contamination by standard bacteriological plating onto 7% sheep blood agar plates. If bacteriological growth is detected the inocula should not be used for parenteral inoculation of any animals until it has been treated with antibiotics or heat.

Addition of antibiotic and/or heat treatment of prepared inocula

Antibiotics are used as the principal method of removing bacterial contamination due to the ease of use and because they are not considered to significantly alter the properties of the inoculum. Ampicillin (500 mg/5 ml reconstituted with distilled water) is the antibiotic used routinely for this purpose.

Ampicillin is added to the thawed inocula samples at the rate of 12.5 µl of antibiotic per ml of inocula. The treated samples are left at room temperature on the laboratory bench overnight. All waste is autoclaved prior to incineration. All samples should then be re-tested for bacterial contamination.

At the VLA, ampicillin (5 mg/4 ml aliquot) has been previously added routinely to suspensions prepared from tongue, rumen, abomasum, duodenum, distal ileum, spiral colon, tonsil, retropharyngeal lymph node and turbinate epithelium where contamination with natural bacterial flora was anticipated [22].

Heat treatment of inocula involves a double application of heat. A blank sample is prepared by filling a bijoux or universal with tap water to the same volume as the inocula samples to be heat-treated and, together with inocula samples, placed in the waterbath. The temperature reached by the samples is tested by checking the water temperature of the blank. When the blank sample temperature reaches 80 °C the treatment of the samples is timed for 15 min. The samples are then placed into an incubator at 37 °C for 1 hour and a further cycle of heat treatment conducted. All the samples should then be re-tested for bacteriological contamination.

Certain tissues have proved acutely toxic to mice on i.c. inoculation (e.g., adrenal) and this effect has been neutralised by prior heat treatment of the inoculum [W. J Hadlow, personal communication].

2.3   Inoculation methods

All staff must be experienced in handling and restraining the species of animals
to be dosed and licensed under specific national legislation for animal experimentation to carry out the procedure.

Inoculation of mice
Inbred mouse strains (RIII, C57Bl6J, IM or VM) of both sexes are inoculated
routinely at 3–4 weeks old. Mice are housed in same sex groups of five as soon as
possible after weaning and each mouse is assigned identification; either by ear
notching (1–5) or by subcutaneous microchip. Twenty mice are used for each
inoculum group. Five individual mouse case cards and a colour coded project
card are clipped to each mouse box lid. The case cards are colour coded
according to sex and detail date of birth, litter and mouse number, inoculum
code, date of inoculation, clinical monitoring, project start date and culling date
(project dependent).

Parenteral exposure
The procedures described here have been used for the bioassay of infectivity in
tissues from cattle with naturally occurring BSE [23], cattle exposed to the BSE
agent [20, 22, 24–26] and pigs experimentally exposed to the BSE agent [21, 27,
28].
    A safety cabinet bench top is covered with disposable absorbent plastic
backed material. Stored inocula, of appropriate dilution are removed from
the freezer and thawed at room temperature.
    The mice are anaesthetised in same sex groups of five using halothane in an
anaesthetic chamber with a gas scavenging system. Once anaesthetised, a
mouse is transferred to the cabinet for inoculation. Each mouse is given an
intracerebral and intraperitoneal inoculation [23], unless the specific project
requires otherwise:

Intracerebral inoculation: A sterile disposable hypodermic syringe and needle
(25G × 16 mm) is used with a disposable polythene sleeve to limit the penetration of the injection to a depth of 3 mm. The transcalvarial injection (20 µl) is
made into the right side of the brain at a point on the skull equidistant between
the ear and the eye.

Intraperitoneal inoculation: A sterile disposable hypodermic syringe and needle
(25 G × 16 mm) is used. The injection (100 µl) is made into the right lower
quadrant of the abdomen. New syringes and needles are used for each inoculum.

Once each mouse is inoculated, it is placed on absorbent paper in a mouse box
in the safety cabinet. When all five mice of each group have been added to the
box each mouse is checked and any haemorrhage from the intracerebral
injection site is removed with cotton wool. The used wool is disposed of as

infected waste. The paper of the mouse boxes is removed a few hours after inoculation and disposed of similarly. The mouse box is removed from the safety cabinet. The mice are checked again 2–3 hours after inoculation and any sick or dead mice are replaced with mice inoculated in an identical manner to those removed. After the post-inoculation period, mice are transferred to an animal room and monitored clinically throughout the course of the study.

Per os exposures

Intragastric exposure: A group of five mice are transferred in their box to a safety cabinet along with a clean box for transfer of each mouse after dosing. Each mouse is removed from its box and restrained by the operator. The inoculum is administered intragastrically by means of a dosing catheter (4.5 FG × 60 mm) attached to a 1 ml syringe. The syringe is filled with inoculum and the catheter attached. A small amount of inoculum is then expelled through the catheter into the inoculum container to ensure that the catheter is filled. The catheter is then introduced into the mouth and passed into the oesophagus and stomach. Inoculum (0.25 ml) is then administered directly into the stomach. The catheter is then removed. The mouse is then transferred to the new box where it is placed on absorbent paper and observed to ensure that there are no adverse reactions to the procedure. Dosing is repeated approximately 4 h later to achieve a final dose of 0.5 ml and the procedure of transfer of the mouse to a new box is repeated. The mice are checked again 2–3 h post challenge. The paper is removed from the box and bagged for incineration and replaced with sawdust. The mice are then transferred to an animal room and are clinically monitored throughout the course of the study.

Feeding: Mice are individually boxed and maintained on a minimum diet throughout the day of challenge. 0.5 ml of inoculum is thoroughly mixed with 1 g of powdered maintenance diet in a sterile glass Petri dish which is then placed in each mouse box in the evening (the period of most active feeding) of the day of challenge. The mice are then observed to ensure that each has consumed all of the feed. The Petri dish is then removed for autoclaving and the mice are given mouse pellets *ad lib*. The mice are then transferred to an animal room and clinically monitored throughout the course of the study.

Inoculation of cattle, sheep and other large animal species

Ideally, for all studies of the transmissibility of a TSE agent test and control animals should be from sources where exposure to TSE agents can be discounted. While this is clearly possible in terms of laboratory rodent studies it has, in the course of the research into BSE and scrapie, only been achieved recently by the use of imported animals from New Zealand and their offspring, reared in isolation. Formerly, sourcing of cattle and sheep for such studies was from UK farms where, as far as possible, it was determined that there was no clinical history of TSE cases and that exposure to meat and bone meal had not occurred.

Dispensing of inocula

The inocula are prepared as previously described and loaded into syringes using drawing up needles (18 G × 40 mm) in the CL3 laboratory and safely transferred to the animal accommodation or surgical facility for use in parenteral or oral dosing. A sheathed drawing up needle is attached to the syringe to prevent leakage and the prepared syringes are placed in individual mini-grip bags and put into a sealable tin or plastic box. Spare inocula are placed in a mini-grip bag and put into a box with sealable lid with spare needles and syringes.

Parenteral exposure

Parenteral routes of inoculation used in cattle are by intracerebral and/or intravenous injection. In sheep, intracerebral and intraperitoneal injections [23] and in pigs, simultaneous intracerebral, intravenous and intraperitoneal injections [21] have been used.

Intracerebral, cattle: The technique was originally developed for the initial studies of the transmissibility of BSE agent to cattle [29–31] and has been used subsequently to assay BSE infectivity by end-point titration in cattle [32] and for the intra species assay of BSE infectivity [26]. Calves from 3.5–6 months old are generally inoculated in groups of five. Protocols for the preparation, anaesthesia and post-operative care of calves used for intracerebral inoculation are not detailed here.

Prior to transport from the animal accommodation to the surgical facility, the hair from the frontal region of the head of the calf is clipped. When restrained on a halter and unloaded at the pre-surgery/anaesthetic area, the identity (ear tag) of each calf is checked against the prepared list of animal numbers and body weights. Following induction of anaesthesia the animal is placed on a stretcher in left lateral recumbency. The stretcher is winched to the operating table and the animal remains on the stretcher throughout the operation. Intubation, for maintenance of closed circuit gaseous anaesthesia, is carried out and the animal is then positioned in sternal recumbency. This is maintained by use of sand bags (plastic covered) with the head secured by straps to maintain it in a stable position.

A formula for determining reproducibility of the inoculation site was devised by a series of measurements conducted post-mortem on the sagittally cut head of calves of the Hereford and Jersey breeds. A generalisation was made from the measurements in the Hereford breed for subsequent use in other breeds, particularly the Holstein/Friesian type. The position of a "Yale" spinal needle (18 G × 9 cm) inserted through a trephine to achieve distribution of inoculum to the required anatomical areas of the brain was noted on a sagittally cut head (Fig. 1).

It was proposed that the inoculum should be injected deep into the brain stem (mesencephalon) and along the needle tract passing through medial parietal cerebral cortex, ensuring installation in multiple neuroanatomical locations, but avoiding major vasculature. The trephine was paramedian in the mid frontal region of the head, at a rostro-caudal level that minimised the depth of the frontal sinus, with the needle inserted at an angle to the midline such that it would traverse the midline ventral to the dorsal sagittal sinus. Approximate

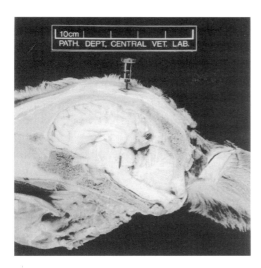

Figure 1 Sagittal section of calf head to show the location of the fully inserted needle for intracerebral inoculation. The needle passes through the cerebral hemisphere and traverses the midline in the rostral midbrain.

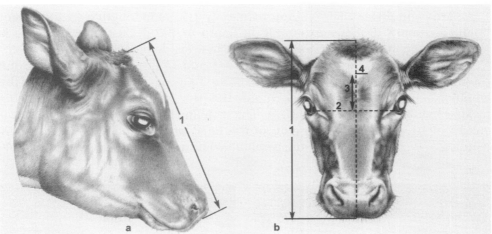

Figure 2 Lateral (a) and frontal (b) diagrams of a calf head showing landmarks and measurements (see text for detail) for use in the formulae to determine the intracerebral inoculation site and depth of insertion of the needle.

measurements determined from the proportions of the head relative to the needle position enabled the application of formulae to calculate the trephine site and depth of penetration of the needle for inoculations (Fig. 2).

Formulae for location of the inoculation site and depth of inoculation needle tract (refer to Fig. 2) are:

(A) Distance (3) from line (2) between medial canthi to inoculation site (midline at 4) = Head length (1) * ÷ 3.44.

(B) Inoculation site from midline (4) = Distance (3) ÷ 6.42.

(C) Depth of inoculation = Head length (1) * ÷ 4.82.

---

* The head length (1) (Fig. 2) is the muzzle to poll measurement obtained using callipers.

The transverse line (2) (Fig. 2b), the line between the medial canthi of the eyes is marked on the clipped forehead of the calf. A colour marker is used depending on hair/skin colour. The distance caudally from the line (2) to the inoculation level on the midline frontal region of the head (3) (Fig. 2b) is calculated [formula (A)] and marked. The trephine site (Fig. 2b) is to the left and lateral to the midline by the measurement (4) which is calculated [formula (B)] and marked.

The depth to which the inoculation ("Yale" spinal) needle is to be inserted into the cranial cavity is calculated by reference to formula (C) and the calculated depth is then subtracted from the length of the inoculation needle (9 cm). This value indicates the length of exposed needle that should remain at the final point of insertion.

The surgical procedure is most efficiently conducted with three operators; two prepared for sterile surgery and a third to prepare the instrumentation.

The inoculation site is prepared with a surgical scrub working from the centre outwards. A disposable sterile drape with a central ovoid aperture is placed over the inoculation site and secured with towel clips.

A parasagittal skin incision, approximately 2.5 cm in length over the inoculation site mark is made extending in depth to the periosteum. Pressure is applied to the incision with a sterile swab, to assist haemostasis.

The sterile trephine instrument (Fig. 3 a and b) is prepared, with careful attention to ensure that the knurled nut in the T piece is secure, the length of the trephine needle exposed is correct (approximately 1.0–1.3 cm) and the securing locking nut on the stop is hand tight. The trephine is made at 90° to the rostro-caudal slope of the frontal region of the head and inclined 25° laterally, by rotating and counter rotating the trephine while applying pressure, until a hole

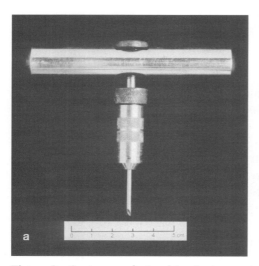
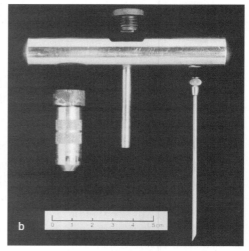

Figure 3 Custom made trephine instrument for the intracerebral inoculation of calves, to provide a cutting edge to a 16 G needle, sheathed with an adjustable stop (to ensure accurate depth of penetration) and a T piece handle; a) instrument assembled for use and b) the separate components.

is made through the skull. To avoid potential damage to underlying blood vessels the operator must be ready to stop applying pressure immediately when penetration of the skull is achieved. Any haemorrhage at this stage on withdrawal of the trephine needle is controlled by applying a swab to the site.

The "Yale" spinal inoculation needle (with stilette *in situ*, but not protruding from the point of the needle) is carefully inserted into the trephine hole maintaining the angles as previously described when trephining, until the calculated length is left exposed, i. e., the needle is introduced to the calculated depth. The needle stilette is then removed and a check is made that there is no blood or cerebrospinal fluid (CSF) back-flow through the needle. If there is backflow of CSF the depth of the needle is adjusted (i. e., it is withdrawn incrementally) until the flow ceases.

The guard and drawing-up needle are removed from the prepared syringe of inoculum and placed directly into a sharps container.

The syringe is agitated to resuspend the inoculum homogenate immediately prior to inoculation. It is then attached firmly to the inoculation needle which is supported and maintained in position. The syringe plunger is withdrawn briefly to check that there is no aspiration of blood or CSF and, if necessary, the needle depth is adjusted as before. Then the syringe is slowly discharged by one operator whilst the other gradually withdraws the needle and syringe. The inocula should be evenly injected along the tract made by the needle through the brain. The needle and attached empty syringe are then carefully removed and transferred directly into a sharps container. To prevent possible personal injury it is important that this removal of the needle is completed by a single operator, the other having cleared from the inoculation site until the needle and syringe have been transferred for disposal.

Pressure is again applied to the site using a sterile swab, the site is swabbed clean and the skin incision sutured. Usually two simple interrupted sutures are used, and the suture needle is subsequently discarded into a sharps container. All soft waste is incinerated.

After suturing, the wound is again cleaned with a dry sterile swab, antibiotic spray is applied and a dressing, comprising a pad of sterile swabs, is applied to the site and secured using elasticated adhesive bandage (7.6 cm width) around the entire head, taking care that respiration is not restricted.

When the procedure has been concluded and the animal disconnected from the anaesthetic equipment it is then carried on the stretcher to the recovery area. The calves are placed in small padded temporary pens. Sternal recumbency is maintained until full consciousness returns. Heaters are used when necessary to prevent chilling. The anaesthetist monitors the animal's recovery at least until removal of the endotracheal tube, ensuring that respiration is normal. Recovery from the anaesthetic to an ability to walk generally takes approximately 3–4 h. All animals are then taken to their allotted accommodation.

Wound bandages are removed after 24 hours and the discarded dressing regarded as infective material and disposed of accordingly.

Intracerebral, sheep: The method for intracerebral inoculation of sheep was developed at another United Kingdom institute [33–35]. The wool/hair overlying the cranium is clipped and the incision site prepared with surgical disinfectant. A 10–20 mm incision in the skin and underlying muscle on the right parasagittal aspect of the cranium is made approximately halfway between a line drawn through the horn scurs and one through the ears. The parietal bone is exposed and a hole is drilled in the skull, using a drill guard to prevent penetration of brain tissue. The inoculum is injected through the drill hole into parietal cerebral cortex (F. Houston, personal communication).

Intracerebral, pigs: The technique was developed for the investigation of the transmissibility of the BSE agent to the pig [21, 27, 28]. Piglets weaned at 1–2 weeks of age are inoculated under gaseous anaesthesia. Each piglet is injected with 0.5 ml inoculum (10^{-1} dilution) into the left cerebral hemisphere by percutaneous, transcalvarial injection. The injection site is located 1 cm lateral to the midline in the frontal region, equidistant from the lateral commisure of the eye and the base of the cranial border of the pinna. The trephine is made with a 16 or 18 gauge hypodermic needle and the inoculation made through a 25 gauge hypodermic needle [21].

Intravenous: Inoculation by this route in cattle and sheep is via the jugular vein and in pigs via the cranial vena cava; standard routes of administration of intravenous veterinary injectable compounds in these species. Animals must be appropriately restrained and when administering the inoculation, care must be taken to avoid back pressure in the syringe (causing explosive spillage and potential contamination of the operator) by using luer lock needles/syringes. Used needles must not be resheathed and must be discarded directly into a sharps container.

Subcutaneous: Although the subcutaneous route of inoculation has not been used in cattle for study of BSE, it is a route that has been used for studies of scrapie in sheep. The injection site is the subcutis of the dorso-lateral aspect of the thorax, immediately caudal to the scapula.

Intraperitoneal: Among large animal species this has only been employed in pigs when the injection is made through the abdominal wall adjacent to the umbilicus with the piglet under general anaesthesia [21].

Per os exposures

Per os exposure to the BSE agent has been used in cattle [20, 22] and sheep [36, 37] by dosing into the oral cavity. *Per os* exposure of pigs [21] to the BSE agent, but with no resultant evidence of transmission, has been conducted by feeding of undiluted brain homogenate prior to access to the normal daily ration. For oral dosing of cattle and sheep, the operator and handler must stand to the side of the animals head in case of regurgitation of the material. The maximum oral dose administered to calves on a single occasion has been 100 g of a pool of BSE-affected brainstem homogenate [22]. This was administered in two successive 50 ml doses using new sterile syringes ensuring that the homogenate was deposited on the base of the tongue at the entrance to the pharynx. Smaller doses of 10 g or less were given as 10 ml amounts in 20 ml syringes. Smaller volumes were made up to 10 ml in sterile saline and dispensed from a 20 ml syringe.

It is assumed that in the immediate post-inoculation period after oral dosing that faecal excretion of inoculum is inevitable with the risk of environmental contamination with infectivity and the possible consequent variable supplementation of the effective individual challenge dose. To prevent this, as far as is possible, a strict post-inoculation husbandry is implemented. This is based on data reviewed on the passage of digesta through the alimentary tract of cattle. In the dairy cow particulate matter is excreted over the period from 12 h to 10 days after ingestion. In smaller ruminants the maximum excretion time is less. Challenged calves are kept on wood shavings, and faeces are removed from the yards several times a day for four days and thereafter once daily for another 10 days after inoculation. This period has later been extended to 28 days to give a wider margin. Similar procedures are used after oral dosing of sheep. A much shorter clearance period would be anticipated for monogastric species, such as the pig.

2.4 Clinical monitoring methods

Mouse studies

Mice are monitored daily during normal husbandry routines. For conventional inbred mouse strains, detailed monitoring [7, 22] commences for all primary inoculations of mice with BSE or scrapie, 250 days post-inoculation and only ceases when a mouse is either killed in extremis, found dead or reaches its terminal kill date (which is predetermined by the nature of the project, inoculum and mouse strain).

The monitoring procedure is carried out once weekly in conjunction with cleaning out, preferably, by the same person on each occasion. The mice should not have been previously disturbed.

Each mouse is examined and a score assigned and recorded before transferring the mouse to a clean box. The assessment needs to be made quickly as the mice become more active on disturbance and it can then be difficult to score accurately. To ensure objectivity the previous week's scoring should not be consulted. Details of affected mice are recorded on the appropriate case card.

The criteria for scoring may require modification according to the project, mouse strain and inoculum (see 3.1).

Cattle, sheep and other large animal species studies

Clinical observations are maintained throughout the studies by a number of approaches. Passive observations, during daily husbandry routines and weekly visits by veterinary staff, are made in order to detect behavioural changes and signs associated with TSE. These features of the clinical presentation are well documented, particularly for BSE in cattle [13, 14, 38–41]. Responses to handling and restraint are assessed during routine weighing and blood sampling at approximately two month intervals from nine months post-inoculation (p.i.). An open field study test has been applied in some studies at approximately two month intervals, beginning 18 months p.i., in an area which permitted free expression of behaviour of individual animals in isolation from their companions [42]. This particularly enabled assessment and detection of abnormalities of kinesis and gait. Behaviour studies have also been made over 24 hour periods by passive observation at approximately three month intervals from 12 months p.i. Clinical neurological examinations are performed at three month intervals and within 7 days of a planned necropsy. A similar, but necessarily modified range of approaches are used for clinical study of sheep with experimental TSEs. Clinical observations of pigs experimentally exposed to TSE agents have, of necessity, because of the difficulties of handling, been largely observational [21]. Observation of behavioural changes is important in this species. Video tapes of the clinical signs of BSE in cattle and scrapie in sheep and goats have been made by several countries and can be obtained on request (Appendix 1).

Routine behavioural observations

The behaviour of cattle and sheep is recorded during normal husbandry routines of weighing, foot trimming and dipping, dipping of sheep and treatments with parasiticides; also during body fluid sampling.

Animals with TSEs are known to display altered behaviour, especially when in unfamiliar environments, when being restrained/handled or when placed in a restraining device. Such behaviour may include apprehension and/or fear towards people. The purpose of recording observations during sampling is to determine any changes in behaviour and to also monitor the progression of such behavioural changes during procedures carried out on a routine basis. Such observations may contribute to define: the normal behaviour of each animal (as when observations are initiated prior to the clinical phase of disease); the time of onset of clinical disease and incubation period and the progression of clinical signs.

The observer watches the animal as it is led from the holding pen and throughout physical restraint procedures, to assess the ease of handling and behaviour.

Weekly observations

Weekly observations have been developed to determine the onset, extent and progression of clinical signs of TSEs in cattle, sheep and pigs that have been exposed to TSE-infected material.

It is important to have easily readable identification marks/tags on animals. Large ear tags are supplemented by photographic evidence of the markings of cattle to be used in the event of loss of tags. The building should be quiet and free from unusual noise and distraction. Animals should not be observed immediately after the pen has been cleaned or fed. The animals should be given time to settle down: at least one hour after disruption, and two hours after feeding. Observations should take place at about the same time of the day, an ideal time being late morning, 2–3 hours post morning feeding (not less than 1 h and not more than 5 h). The observer should position themselves at the same site each time, to ensure consistency of the results. The animal's activity when the pen is first approached is recorded. The animal's reaction to the observer's movements, without touching the animal, is tested. Typically a pen of five animals is observed for a minimum of a 15 min period, allowing about 3 min per animal. Further descriptions of any other clinical signs or unusual findings should be recorded.

Neurological examination

A neurological examination is conducted to assess an animal's mental state, certain central nervous system functions and to check the integrity of certain nervous pathways [43].

Neurological examinations are important to rule out existing neurological deficits in calves and lambs before their allocation to specific TSE projects and therefore determine their suitability for the study, to assess the clinical status of project animals over time, and to diagnose clinically TSE or other neurological diseases in project animals.

The neurological examination is carried out by a veterinary clinician along with a member of scientific staff. Details of the examination are recorded systematically on the examination form appropriate to species (see Appendices 2 and 3). Anomalies of the optic fundus or nerve are drawn on the Optic Fundus Record Form (see Appendix 4).

Specific behavioural studies on cattle

Behavioural observations of cattle have been used in some studies in the past to assess general behaviour such as eating, ruminating or idling and to determine the clinical status/monitor the progression of clinical disease [22, 40, 41]. The night before the test (if duration of study is 24 h) all lights in the animal building are left on to minimise disturbance to the animals during the period of observation. All animals to be observed are numbered with different coloured labels, on the flanks and on both sides of the rump, early on the morning of the study. These animals are then left for at least 2 h to settle down following labelling. In animal buildings where there is no observation gantry, a viewing tower is set up to allow each animal to be seen clearly. Observers should enter the animal area as quietly as possible, to cause limited disturbance to the animals. The environmental conditions should be noted, recording the temperature inside the building, the presence of any flies and any noise conditions throughout the study (specifying particularly the time of any extreme noise conditions).

The behaviour of each animal is recorded at either 5 or 10 min intervals over the 24 h period using abbreviations, i. e., SI (Standing Idle), SR (Standing Ruminating), SE (Standing Eating), SD (Standing Drinking), LI (Lying Idle), LR (Lying Ruminating), LE (Lying Eating) and LD (Lying Drinking). Each feature of behaviour is then expressed as a percentage frequency of occurrence over the observation period [40].

Where possible, any animal displaying excessive behaviour/any unusual behaviour should be recorded on video tape.

Change-over of shift staff should be done quickly and quietly, to avoid disturbing the animals.

Guidelines for the interpretation of the clinical approaches used in studies of cattle inoculated with BSE are given below (3.1).

2.5 Necropsy procedures

Here we consider only those procedures involved specifically in the sampling of tissues for TSE transmission studies involving mice, cattle and sheep although in generalities they will apply also to other species. Readers are referred to standard texts for post-mortem dissection methods for each species.

Tissues are collected at necropsy for a variety of purposes according to various post-mortem protocols. In general, tissues will be processed, immediately following sampling, in four ways: fixed for histological examination; aseptic collection of fresh tissues for freezing at −80 °C, snap frozen in liquid nitrogen or fresh frozen on solid carbon dioxide.

Mouse studies
Mouse necropsy procedures are conducted in a Class 1 safety cabinet.

The majority of protocols will require the following, or permutations thereof: 1) non-aseptic removal of entire brain into phosphate buffered neutral, 10% formalin for histological examination; or 2) aseptic removal of entire brain onto sterile aluminium foil, dorsal surface uppermost, for dissection. For the latter a parasagittal cut is made, just to the right of the midline with a sterile scalpel, to remove approximately one third of the brain mass into a pot for fresh freezing at −80 °C for potential sub-passage. The remainder of the brain is placed into formalin for histological examination. For certain experimental protocols and/ or for the monitoring of intercurrent disease, various viscera may be required to be sampled.

Cattle and sheep studies
The necropsy of cattle and sheep is conducted in an appropriate large animal post-mortem facility.

Tissue should be removed in the order of their preparation: snap-freezing, aseptic fresh frozen and then histological examination, unless the protocol requires otherwise.

Snap freezing of samples: Tissue is cut into 2 mm cubes/blocks and placed into cryovials which are then immersed in liquid nitrogen until frozen. After snap freezing the samples are placed on solid carbon dioxide (cardice) until they are transferred to a freezer (-80 °C).

Where the protocol specifies aseptic removal/sampling, two operators are necessary. The tissue is exposed by the first operator taking care not to touch tissue with non-sterile forceps or scalpel. A second operator wearing clean gloves and using sterile forceps and scalpel, removes the tissue avoiding contact with any of the surrounding tissue, and places it into a sterile Petri dish. The Petri dish is moved to a clean area of the post-mortem room and the tissue is trimmed of all adherent connective tissue. The tissue is then aliquoted as specified in the protocol and subsequently placed at –80 °C.

Tissues for histological examination are dissected from adjacent connective tissue and immersed in the relevant fixative. Tissues, either dissected/blocked or as whole organs (e. g., brain), are placed into a volume of fixative 10 times that of the tissue volume. The technical procedures concerned with collection, fixation and histological processing of brain described previously for diagnosis of BSE in cattle [44–46] are, in general, also used for the microscopic study of central nervous tissues in other animal models. Appropriate modifications are made for the size of the brain from different species, in relation to fixation in formalin and for processing schedules and staining methods.

Tissue blocks other than brain and spinal cord must not exceed 1 cm thickness for good primary fixation and some require specific pre-treatments. Once dissected, peripheral nerves are placed under slight tension on cardboard, left for 5–10 min to adhere, the cardboard is trimmed leaving about 5 mm

Figure 4 Muscle clamp used to aid sampling of skeletal muscle of cattle and sheep prior to fixation for histological examination.

around the tissue and then placed in a labelled honey jar containing 10% formol saline.

While *in situ* skeletal muscles are placed under slight tension and two cuts, 0.5 cm apart and 10 cm long, are made parallel to the direction of the fibres. With the muscle under tension a muscle clamp (Fig. 4) is placed onto the dissected strip of muscle. After ensuring the clamp is attached firmly the strip of muscle tissue is removed with the clamp by severing the muscle strip from the remaining muscle. The clamped muscle strip is placed in warm physiological saline (~37 ºC) for 15 min and then transferred to buffered formalin.

Microscopic pathology

For the morphological examination of histological sections the standard staining method employed from the earliest examinations of the brains of cattle with BSE has been a modified haematoxylin and eosin (HE) technique [13]. By reducing differentiation of the haematoxylin, the section retains much of this blue dye giving a high contrast between stainable and non-stainable elements, allowing ease of identification of vacuolar changes. This method is applied to histological study of the central nervous system of all species discussed in this chapter. Other routine tinctorial methods include the luxol fast blue, cresyl violet and congo red techniques.

The principal immunohistochemical method used to study the microscopic changes is that directed to the visualisation of PrP^{Sc}. Polyclonal antibodies and monoclonal antibodies to C and N terminus regions of the protein are used, together with a range of epitope demasking and signal amplification techniques, on central nervous and peripheral tissues. Protocols are optimised specifically for use with each antibody.

3   Results and discussion

3.1   Data interpretation and analysis

Clinical

Clinical findings are utilised to indicate the onset and progression of disease and to thereby define incubation period. They may also contribute to define a disease phenotype on transmission to the same species or on transmission across a species barrier in biological strain typing (see Chapter 10).

Guidelines for the interpretation of clinical observations with respect to determining a clinical endpoint for mice and cattle inoculated with the BSE agent are as follows:

Scoring of mice (e. g., RIII inoculated with BSE agent)

Adapted from "Clinical Criteria Used to Define Incubation Period" [7].

1 = Normal mouse

2 = First signs of BSE

Must show major sign marked * OR two other signs.

* Waddling gait

  Rough coat

  Dullness

  Hyper-responsive

  Incontinence

  Paresis with splaying of hind limbs and loss of normal spinal curvature.

+ = Near to terminal BSE

Must show two major signs marked *

* Marked gait abnormality

* Eye discharge and dullness about head

* Weight loss

* Paresis with splaying of hind limbs and loss of normal spinal curvature.

  Incontinence

  Rough coat

  Hunched

Mice are culled if they:

(1) Score + on three consecutive weeks

OR (2) Show advanced signs and death is considered to be imminent.

ASSESSMENT OF CLINICAL STATUS OF CATTLE IN BSE STUDIES

The onset of clinical signs of BSE is slow and insidious in most experimental cases, and the BSE clinical status of cattle may be described as either clinically negative/ no apparent signs of BSE (Status 0), possible BSE (Status I), probable but early BSE (Status II), and definite BSE (Status III). Mild clinical signs compatible with an early BSE but not specific have been observed intermittently and insidiously over long periods in some cattle challenged with BSE. The classification of an animal into one of the three clinical categories is somewhat subjective and when progression takes place, the exact time of transition from one status to the next is not easily defined in all animals.

Possible BSE case (Status I)

There is occasional display of clinical signs consistent with BSE but of mild intensity or severity. These possible signs may include behavioural changes, such as mild aversion to attendants, loss of confidence in approaching observers, a changed relationship with other animals in the group, a subtle change of behaviour when confronted with a new environment, tooth grinding, nose licking/head rubbing of low intensity. Other signs may be exaggerated responses to tactile, auditory or visual stimuli or subtle gait abnormalities. These clinical signs are not necessarily caused by a neurological dysfunction and may also be present in a small percentage of healthy control cattle. They may be caused by other factors, such as pruritus resulting from ectoparasitism, aversion arising from previous adverse experiences (i. e., sequential bleeding), visual impairment or age-related degenerative joint diseases. The display of signs of different categories (subtle changes in behaviour, sensation and locomotion) at the same time is more likely to reflect an abnormality related to BSE.

Probable BSE case (Status II)

Clinical signs of mild intensity or severity are displayed regularly but not always, with or without more severe signs occasionally. Clinical signs have been observed over at least 3 weeks in a 5 weeks period (no less than three assessments). The clinical changes can be discriminated from normal behaviours on the basis of the regularity and novelty of their expression in a specific animal or of their disease-specific nature. Neurological abnormalities and disease progressiveness are usually not obvious or permanent.

Definite BSE case (Status III)

The clinical signs of BSE are constant and usually include unequivocal signs of a neurological disease. This stage usually follows a period when probable signs of BSE were displayed regularly and with an increasing severity. The progressive nature of the disease has been documented over not less than three clinical assessments done on a frequency to be determined by the type and rate of progression of clinical signs.

In most cases, a possible status has been observed first in the clinical progression of BSE. In some animals, this stage was not observed, possibly because the onset of clinical signs was truly more sudden than in most cases or because individual animals were not observed with scrutiny.

Mouse bioassay (qualitative)

The primary purpose of the qualitative bioassay is to obtain a positive or negative diagnosis in relation to spongiform encephalopathy (SE) from the brightfield light microscopical examination of the standard HE stained sections (Fig. 5) of the mouse brains obtained from the bioassay study. In most studies this is supplemented by an examination of the distribution profile of any SE vacuolation present. Nine grey matter areas [47] and three white matter areas

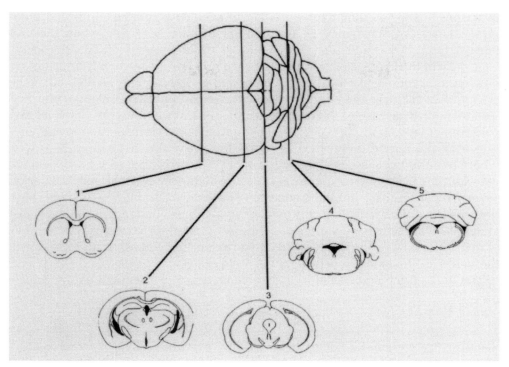

Figure 5 Standard levels of mouse brain sectioned for microscopical examination in the qualitative and quantitative assays. Tissue blocks 1–4 are sampled rostral to the corresponding transverse cuts and tissue block 5 is taken from the medulla/cerebellum caudal to the most caudal cut.

are scored [48]. Where there is significant vacuolation identified in other areas this is also recorded. This procedure, "lesion profiling", is an integral part of biological strain typing, which is dealt with in detail in Chapter 10.

Most of the microscopical assessment is achieved at magnifications of X25–X100. Experience is required in the differentiation of SE vacuolation from age, viral or artefactually induced vacuoles.

In the bioassay, transmission is defined as occurrence of histopathological evidence of SE, or, where applied, immunohistochemical presence of PrPSc in the brains of the mice. The use of immunohistochemical detection of PrPSc was introduced to the standard protocol for assessment of selected bioassay results to provide improved specificity and sensitivity of detection of BSE transmission to mice [25] and enabled interpretation of inconclusive histopathological results, which arise particularly when age-associated vacuolar changes, prominent in old female mice, may closely resemble TSE specific vacuolation [49].

In qualitative bioassays of tissue infectivity using wild type mice, results are usually tabulated according to inoculated tissue, to provide information on the incidence of mice developing disease and mean incubation period or survival period with standard error of mean. The incidence of mice succumbing to

disease is usually defined by the number of mice with confirmed histopathological changes of SE divided by the number of mice surviving when the first
positive mouse was confirmed. This definition excludes incidental deaths occurring early in the incubation period and provides a more statistically appropriate
denominator for disease incidence. Incubation period can only be defined by the
standardised clinical endpoint described previously, but because of the impreci-
sion of clinical assessment alone, especially in assays of low titre material and
consequent long incubation periods, histopathological confirmation in conjunction with the period of survival provides the proxy for incubation. However,
where immunohistochemical detection of PrP^{Sc} alone is used to define a positive
case the time from inoculation to death can be defined only as the survival period
since in most, if not all, animal models of TSEs PrP^{Sc} accumulation precedes both
occurrence of histopathologically evident spongiform change and clinical signs.

For assays in which no mice develop disease it is useful to express the number
of mice that survived to the terminal kill/number inoculated, as an indicator of
the power of the assay in proving a negative result.

The limit of detectability of infectivity in an assay is discussed below.

Mouse bioassay (quantitative)
The concentration of infectious TSE agent in tissue is determined by bioassay
usually using endpoint titration or, sometimes, although regarded less accurate,
by incubation time assay [50]. The experimental models that have most often
been used for such assays are inbred strains of mice. While the most practical
bioassay model, wild type mice are likely to provide an underestimate of the
concentration of agent in, for example, sheep or cattle tissues, because of the
effect of the species barrier. Titration data are usually expressed in the form of
$\log_{10}$ ID$_{50}$ units according to species and route(s) of inoculation [51, 52]. One
mean infective (or lethal dose) (ID$_{50}$) is defined as the amount of infectivity that
will transmit disease to half of a group of inoculated animals. If 1 ml volumes of
successive 10-fold dilutions of a specimen are inoculated into a total of 20 mice
per dilution group, one ID$_{50}$ would be present in the dilution that transmits
disease to 50% (10/20) of the inoculated animals. Titrations often show transmission rates of about 100%, 50%, and 0% in successive 10-fold dilutions and
providing the last, the limiting dilution, is determined, the survivorship can be
used to calculate the units of ID$_{50}$ in the original undiluted inoculum volume.
This is usually corrected to express the units of ID$_{50}$/g of the tissue.

It must be stressed that experience of endpoint titration and incubation
period assays in laboratory rodents suggests that the assays are most reproducible when examining infectivity of a strain of agent adapted or, ideally,
cloned, in the assay species and where the concentration of infectivity is high.
The inter-laboratory reproducibility of endpoint titration assays, at least with
ME7 scrapie has been demonstrated in C57Bl mice [53]. The recent use of
mouse titrations of infectivity for cattle and sheep tissues is in contrast to this
since the titrations have been conducted across a species barrier on primary
inoculation from cattle or sheep.

There are a number of factors which affect the efficiency of infectivity assays. Route(s) of inoculation affects infectivity titre and dose response curves [54]. The volume(s) of inoculum injected also affects the sensitivity of the assay. In practice, the most efficient route(s) of inoculation is selected to perform the assay. In previous assays of scrapie agents in mice the calculated limit of detectability of scrapie infectivity by the intracerebral (i.c.) inoculation of mice was approximately $10^{2.0}$ $\log_{10}$ mouse i.c. ID_{50}/g of tissue [55] with a volume of inoculum of 30 μL. Clearly, the volume of inoculum that can be injected intracerebrally in a mouse is limited. In the mouse assays of infectivity of tissues from cases of naturally affected cattle with BSE [23 and H. Fraser, personal communication], a combination of i.c. and intraperitoneal (i.p.) injections was used with a total volume of inoculum of 120 μL, giving a limit of detectability of $10^{1.4}$ mouse i.c./i.p. ID_{50}/g [56]. Endpoint titration of BSE on primary pass to RIII and C57Bl inbred mouse strains gives closely similar values for infectivity [57].

The relationship between incubation period and titre has been questioned by many authors [58], casting doubt on the validity of estimating titre by incubation period assays. Certain physical and chemical treatments of scrapie affected mouse brain inocula may alter incubation relative to titre, giving a discrepancy between endpoint titration and incubation period assay of about 10^{1}–10^{2} ID_{50} [58]. In an analysis of over 100 scrapie infectivity titrations in mice [59] a linear rise in mean incubation period with logarithmic decreasing dose was substantiated, but variability in incubation period rose linearly as dose decreased. Thus estimation of titre from dose response curves is less accurate at low doses. Comparisons of titrations of infectivity in spleen and brain in a hamster model of scrapie suggest also that the relationship of inoculum dose to incubation may differ between organs [60].

Titration of infectivity in TSEs has been performed largely on central nervous system tissue, notably brain.

3.2 Applications

Many of the uses of the methods employed for the study of TSEs in animal models have been discussed in the preceding sections; here we summarise briefly the extent of such applications. Animal inoculation studies with TSE agents, particularly in mice [61], have been used to explore transmissibility, host range, species barrier effects, routes of exposure, susceptibility (especially the influence of host genetic factors), assay of infectivity, properties of the causal agent, inactivation strategies, biological strain typing, pathogenesis and therapeutic approaches. Transgenic manipulation of the PrP gene has enabled investigation of the specific effects of this gene on susceptibility and has improved understanding of species barrier effects [62–66]. The PrP "knockout", partial deletion and chimeric PrP gene constructs and constructs in which the species barrier has been eliminated are among many established and emerging transgenic animal model strategies. Transgenic mouse transmissions are indicating that the basis of "strain" may be in the conformational templating properties of PrP^{Sc} on the normal form of the prion protein (PrP^C). Ectopic, cell directed, expression of PrP is also providing information on the behaviour of PrP.

While these Tg applications offer much in terms of studying *in situ* properties of PrP, in the context of the use of animal models in the study of TSE, they are distanced from an understanding of the naturally occurring TSEs and risk assessments of their impact on human and animal populations, suggesting that many of the conventional methods described in this Chapter may remain important to TSE research for the foreseeable future.

Acknowledgements

The authors gratefully acknowledge the many co-workers at VLA, past and present who have contributed to this Chapter. We thank Mrs L.P. Cooper for word processing. We are indebted to the Institute for Animal Health, Neuro-pathogenesis Unit, Edinburgh, for cooperation over many years in providing assistance and training to staff to ensure faithful replication of the approaches used in transmission studies with inbred mice strains which they have evolved over several decades.

Appendix 1: Videos of the clinical signs of BSE in cattle and scrapie in sheep

1. ADAS/MAFF (1988) Bovine Spongiform Encephalopathy. (7 min 40 s)
2. MAFF (1993) The Neurological Clinical signs of BSE, produced by Gerald Wells. (15 min 28 s).
3. Clinic of Ruminant and Equine Medicine, University of Zurich (1997). Clinical Findings in Cows with bovine spongiform encephalopathy (BSE), produced by U Braun, N Pusterla, E Schicker (20 min).
4. Department of Agriculture and Food, Ireland (1999). Clinical Signs of BSE, produced by Dr Maire McElroy, Veterinary Research Laboratory, Abbotstown (13 min) Farm TV Production.
5. ENVA France (2000). Aspects cliniques de l'ESB et de la tremblante des petits ruminants. Diagnostic différential de l'ESB, produced by Jeanne Brugère-Picoux (40 min).
6. MAFF (date unknown). Scrapie: Clinical signs in sheep and goats. Assembled by members of the European Union Scientific Veterinary Committee. Produced by MAFF Publicity Video Unit, London, UK for the European Community (18 min 30 s).

Appendix 2: Neurological examination of cattle

Animal No:	Farm:	Pen:	
Date:			On database: On Study Diary:
Clinician:	Reason for Examination:		
	Rec. by:		

if normal: tick (✓) - if abnormal: circle (when listed) or describe in detail and place a "＊"in the margin, detail in #43 if insufficient space in box – if test not performed: cross out (×), and always indicate why if non-performance is due to the animal's reaction

ANIMAL FREE (also 33.)

1. **Posture** (head, neck, limbs, back)			
2. **Walking** (amount/willingness)			
3. **Turning**			
4. **Running** (amount/willingness)	Trot:	Gallop:	
OVERALL GAIT 5. **Stiff/Lame**	No / Yes, describe:		
6. **Neurological**	No / May be / Yes, describe:		
7. **Other on gait**	No / Yes, describe:		
8. **Slipping**/ 9. **Falling** (describe if yes)	No / Yes:	No / Yes:	
10. **Obstacle** (device: _____)			
11. **Acceptance of crush** (going in)			

ANIMAL IN CRUSH (also 34.)	Symmetry	Left	Right
12. **Sclera visible** (amount/pigmented?)			
13. **Eye position** (strabismus?)			
14. **Eyelid position** (ptosis?)			
15. **Third eyelids** (position/colour)			
16. **Nose** (sym. & movements/breath)			
17. **Menace response**			
18. **Ears** (position and reaction to touch)			
19. **Blink** (lateral and medial canthus)			
20. **Nose** (reaction to touch)			
21. **Lips** (sym./reaction to touch: *smile)*			
22. **Eye movements**			
23. **Sweat beads on muzzle**			
24. **Salivation** (✓, ↑, or ↑↑)			
25. **Jaw position / Tongue tone**	/		

HEAD RESTRAINED/ HALTER (also 35.)	Symmetry	Left	Right
26. **Optic nerve/fundus**			
27. **Light reaction** (direct and consensual)			
28. **Corneal reflex**			
29. **Cutaneous trunci and neck prick**	CT	NP	SP
30. **Lordosis/kyphosis**			
31. **Tail tone/anal tone**			

OVERALL ASSESSMENT

32. **Mental status** ✓(normal), dull, depressed, "hyper", etc.	
33. **Behaviour & reactivity free** ✓(normal), excited, playful, fearful, nervous, friendly, boisterous, dangerous, "hyper", active, quiet etc.	
34. **Behaviour in crush** ✓(normal), quiet, restless, agitated, agitated 1st then settled down, never settled down, frantic, etc.	
35. **Behaviour**/head restraint & head tests	HR HT
36. **Clipboard test**	
37. **Bang test**/38. **Flash test**	37. H Clap 38.
39. **Flexible stick test**	
40. **Tremors**	No / Yes, describe:

41. **Body condition:**

42. **Extraneural findings:**

43. **Other/more on ...***(#)***:**
                44. <u>**Status:**</u>

       Animal No: **Date:**

GENERAL EXAMINATION

45. **Temperature:**   49. **Dehydrated?**

46. **Heart rate:**   50. **Mucous membranes:**

47. **Ruminal Contractions:**   51. **Additional Findings:**

48. **Lymph nodes**:   52. **Behaviour in pen prior to Exam:**

     TO DO:

ACTION	DATE COMPLETED	RESULT
Bloods taken:		
Urine taken:		
Skin scraping (location):		
Video of (describe):		
Still photograph of (describe):		
Other :		

Appendix 3: Neurological examination of sheep

STUDY: **DATE:**	On Database:
ANIMAL NO: **FARM OF ORIGIN:**	On Study Diary:
BREED: **SEX:** **FARM:** **BLOCK:** **PEN:**	
CLINICIAN: **RECORDED BY:** **REASON FOR EXAMINATION:**	

(✓ if normal , × if not able to observe, describe in detail if abnormal)

1. **Posture when approached**	Standing Sitting sternal other:
2. **Posture/Head Carriage**	
3. **Abnormal Movements/ Tremors/Fasciculations**	**None** / **Yes** (describe):
4. **Mental Status**	
5. **Behaviour**	**Resting (left alone):** **Approached:** **Handled:**
6. **Hand clap**	

FACE/NECK	**Symmetry** (tick if all normal, or describe)	**Left**	**Right**
7. **Optic nerve/Ocular fundus**			
8. **Light reaction**			
9. **Eye position** (strabismus?)			
10. **Eyelids**/11. **Third Eyelids**			
12. **Menace response**			
13. **Corneal reflex**			
14. **Ears** (position/reaction to touch)			
15. **Blink** (lateral/medial canthus)			
16. **Lips** (sym./reaction to touch)			
17. **Nose** (movements/touch)			
18. **Eye movements**			
19. **Blinding to side**			
20. **Salivation, dribbling**			
21. **Jaw position/tone**		**Tongue tone**	
22. **Swallow** (on palpation)			

23. **Scratch/Nibble Test**	**Negative** / **Inconclusive** / **Inconsistent** / **Positive** / **Unable to assess**					
Magnitude/duration/onset: Weak Clear Exaggerated Delayed Prolonged **Other / details:**	**Area / Reaction**	**Nibble/Lick**	**Head mov**	**Body mov**	**Other react**	**Sensitive**
	Cranial thoracic area					
	Middle thoracic area					
	Thoracolumbar					
	Lumbar area					
	Sacral area					
	Base of the tail					
	Shoulder					
	Side thorax					
	Side abdomen					
	Thigh					

24. **Cutaneous trunci**	
25. **Body condition**	**Body condition score:**
26. **Anal and tail tone**	
27. **Withdrawal reflex**	
28. **Righting reaction** (symmetry)	
29. **Behaviour blindfolded**	

GAIT	
30. **Walking** (amount/willingness)	
31. **Running** (amount/willingness)	
32. **Slipping**	**No** / **Yes** (# of times, describe):
33. **Falling**	**No** / **Yes** (# of times, describe):
34. **Collides into gates/ walls**	**No** / **Yes** (# of times, describe):

STUDY NUMBER:　　　　　　　　　　　　　　　　　　**ANIMAL NO:**

　　　　　　　　DATE:

OVERALL GAIT:	
35. **Stiff/ lame**	**No** / **Yes** – describe:
36. **Neurological**	**No** / **May be** / **Yes** – describe:
37. **Other on gait**	**No** / **Yes** – describe:
52. **Limb adduction - abduction**	
53. **Side - push**	

38. **Other On Behaviour/Activity** (tooth grinding, rubbing):

39. **Skin lesions: No** / **Yes** – list areas and types/ *draw below*

40. **Wool – hair loss: No** / **Yes** – list areas and types/ *draw below*

41. **Other Fleece change(s): No** / **Yes** – list areas and types:
also draw below (soiling/ changed fleece colour/ quality)

R　　　L

42. **Cud Soiling: No** / **Yes** : nostrils/chin/other:　　　**Cud dribbling: No** / **Yes** : episodic/permanent, other:

43. **Other / more on #:**

GENERAL EXAMINATION (*go back and complete 1–6, then verify that all boxes have entries after this step*)
45. **Temperature:**　　　　　　　48. **Lymph nodes:**　　　　　　　51. **Additional findings:**

46. **Heart rate:**　　　　　　　49. **Dehydrated?**

47. **Ruminal contractions:**　　　　50. **Mucous membranes:**

44. **STATUS WITH REGARDS TO TSE & OTHER DISEASES:** - no evidence of TSE ? Normal Animal / Other abnormality (ies) - suspect TSE → severity of case: early – established – advanced 　　　　　　　　→ list suggestive signs (by their # if many): - inconclusive with regards to TSE , detail: - signs suggesting another neurological illness - signs suggesting another (non neurological) illness **No** / **Yes** – Detail:

HISTORY:

ACTION	DATE COMPLETED	RESULT
Still photo: **No** / **Yes** (describe):		
Video: **No** / **Yes** (describe):		
Skin scraping: **No** / **Yes** (describe):		
Other (describe):		

PG Number:　　　　　　　　　　　　　　　　　**Date of post mortem:**

Appendix 4: Fundus examination

Project: _____

Animal: _____

Date: _____

OS   OD

Ocular Fundus

References

1   Wilkinson L (1992) From transmissibility of rabies and glanders to the bacteridium of anthrax 1800–70. In: L Wilkinson: *Animals & Disease. An introduction to the history of comparative medicine.* Cambridge University Press, Cambridge 115–130

2   Cuillé J, Chelle PL (1936) La maladie dite tremblante du mouton est-elle inoculable? *Compte rendu d l'Académie des Sciences* 203: 1552–1554

3   Hadlow WJ (1959) Scrapie and kuru. *Lancet* 2: 289–290

4   Chandler RL (1961) Encephalopathy in mice produced by inoculation with scrapie brain material. *Lancet* 1: 1378–1379

5   Gajdusek DC, Gibbs CJ Jr, Alpers M (1966) Experimental transmission of a kuru-like syndrome to chimpanzees. *Nature* 209: 794–796

6   Gibbs CJ Jr, Gajdusek DC, Asher DM et al. (1968) Creutzfeldt-Jakob disease (spongiform encephalopathy): transmission to the chimpanzee. *Science* 161: 388–389

7   Dickinson AG, Meikle VMH, Fraser HG (1968) Identification of a gene which controls the incubation period of some strains of scrapie agent in mice. *J Comp Pathol* 78: 293–299

8   Prusiner SB, Groth DF, Cochran SP et al. (1980) Molecular properties, partial purification, and assay by incubation period measurements of the hamster scrapie agent. *Biochemistry* 19: 4883–4891

9   Prusiner SB, Cochran SP, Groth DF et al. (1982) Measurement of the scrapie agent using an incubation time interval assay. *Ann Neurol* 11: 353–358

10 Hadlow WJ, Race RE, Kennedy RC et al. (1979) Natural infection of sheep with scrapie virus. In: SB Prusiner, WJ Hadlow (eds): *Slow Transmissible Diseases of the Nervous System.* Academic Press, New York, 331–356

11 Hadlow WJ, Kennedy RC, Race RE, Eklund CM (1980) Virologic and neurohistologic findings in dairy goats affected with natural scrapie. *Vet Pathol* 17:187–199

12 Hadlow WJ, Kennedy RC, Race RE (1982) Natural infection of Suffolk sheep with scrapie virus. *J Infect Dis* 146: 657–664

13 Wells GA, Scott AC, Johnson CT et al. (1987) A novel progressive spongiform encephalopathy in cattle. *Vet Rec* 121: 419–420

14 Wilesmith JW, Wells GAH, Cranwell MP, Ryan JBM (1988) Bovine spongiform encephalopathy: epidemiological studies. *Vet Rec* 123: 638–644

15 Wilesmith JW, Ryan JB, Atkinson MJ (1991) Bovine spongiform encephalopathy: epidemiological studies on the origin. *Vet Rec* 128: 199–203

16 ILSI Europe Report (2003) Transmissible Spongiform Encephalopathy as a Zoonotic Disease. Report prepared under the responsibility of the International Life Sciences Institute (ILSI) Europe Emerging Pathogen Task Force with the Endorsement of the International Forum for TSE and Food Safety (TAFS). ISLI Europe, Brussels

17 Department of Health (2003) Transmissible spongiform encephalopathy agents: safe working and prevention of infection. Published 15 December 2003. http://www.dh.gov.uk/policyandguidance/healthandsocialcaretopics/cjd (Accessed June 07, 2004)

18 Hope J, Reekie LJ, Hunter N et al. (1988) Fibrils from brains of cows with new cattle disease contain scrapie associated protein. *Nature* 336: 390–392

19 Prusiner SB, Fuzi M, Scott M et al. (1993) Immunologic and molecular biologic studies of prion proteins in bovine spongiform encephalopathy. *J Infect Dis* 167: 602–613

20 Wells GA, Hawkins SA, Green RB et al. (1998) Preliminary observations on the pathogenesis of experimental bovine spongiform encephalopathy (BSE): an update. *Vet Rec* 142: 103–106

21 Wells GA, Hawkins SA, Austin AR et al. (2003) Studies of the transmissibility of the agent of bovine spongiform encephalopathy to pigs. *J Gen Virol* 84: 1021–1031

22 Wells GAH, Hawkins SAC, Green RB et al. (1996) Preliminary observation on the pathogenesis of experimental bovine spongiform encephalopathy. In: CJ Gibbs Jr (ed): *Bovine spongiform encephalopathy: the BSE dilemma*. Serono Symposia USA. Springer-Verlag, New York, 28–44

23 Fraser H, Foster J (1994) Transmission to mice, sheep and goats and bioassay of bovine tissues. In: R Bradley, B Marchant (eds): *Transmissible Spongiform Encephalopathies*. A Consultation on BSE with the Scientific Veterinary Committee of the Commission of the European Communities held in Brussels, 14–15 September 1993. Document VI/4131/94-EN. European Commission, Agriculture, Brussels, 145–159

24 Wells GA, Dawson M, Hawkins SA et al. (1994) Infectivity in the ileum of cattle challenged orally with bovine spongiform encephalopathy. *Vet Rec* 135: 40–41

25 Wells GA, Hawkins SA, Green RB et al. (1999) Limited detection of sternal bone marrow infectivity in the clinical phase of experimental bovine spongiform encephalopathy (BSE). *Vet Rec* 144: 292–294

26 Wells GAH, Spiropoulos J, Hawkins SAC, Ryder SJ (2004) Pathogenesis of experimental bovine spongiform encephalopathy (BSE): preclinical infectivity in tonsil and observations on the distribution of lingual tonsil in slaughtered cattle. *Vet Rec*: in press

27 Dawson M, Wells GAH, Parker BNJ, Scott AC (1990) Primary parenteral transmission of bovine spongiform encephalopathy to the pig. *Vet Rec* 127: 338–339

28 Ryder SJ, Hawkins SAC, Dawson M, Wells GAH (2000) The neuropathology of experimental bovine spongiform encephalopathy in the pig. *J Comp Pathol* 122: 131–143

29 Dawson M, Wells GAH, Parker BNJ (1990) Preliminary evidence of the ex-

perimental transmissibility of bovine spongiform encephalopathy to cattle. *Vet Rec* 126: 112–113

30 Dawson M, Wells GAH, Parker BNJ, Scott AC (1991) Transmission studies of BSE in cattle, hamsters, pigs and domestic fowl. In: R Bradley, M Savey, B Marchant (eds): *Subacute Spongiform Encephalopathies*. Commission of the European Communities. Kluwer Academic Publishers, Dordrecht, 25–32

31 Dawson M, Wells GAH, Parker BNJ et al. (1994) Transmission studies of BSE in cattle, pigs and domestic fowl. In: R Bradley, B Marchant (eds): *Transmissible Spongiform Encephalopathies*. A Consultation on BSE with the Scientific Veterinary Committee of the Commission of the European Communities held in Brussels, 14–15 September 1993. Document VI/4131/94-EN. European Commission, Agriculture, Brussels, 161–167

32 Hawkins S, Wells G, Austin A et al. (2000) Comparative efficiencies of the bioassay of BSE infectivity in cattle and mice. In: *Proceedings of the Cambridge Healthtech Institute's 2nd International Transmissible Spongiform Encephalopathies Conference*, 2–3 October 2000, Alexandria, Virginia, USA

33 Foster JD, Hope J, Fraser H (1993) Transmission of bovine spongiform encephalopathy to sheep and goats. *Vet Rec* 133: 339–341

34 Foster JD, Bruce M, McConnell I et al. (1996) Detection of BSE infectivity in brain and spleen of experimentally infected sheep. *Vet Rec* 138: 546–548

35 Foster JD, Parnham D, Chong A et al. (2001) Clinical signs, histopathology and genetics of experimental transmission of BSE and natural scrapie to sheep and goats. *Vet Rec* 148: 165–171

36 Jeffrey M, Ryder S, Martin S et al. (2001) Oral inoculation of sheep with the agent of bovine spongiform encephalopathy (BSE). 1. Onset and distribution of disease-specific PrP accumulation in brain and viscera. *J Comp Pathol* 124: 280–289

37 Foster JD, Parnham DW, Hunter N, Bruce M (2001) Distribution of the prion

protein in sheep terminally affected with
BSE following experimental oral transmission. *J Gen Virol* 82: 2319–2326

38 Wells GAH (1989) Bovine spongiform
encephalopathy. In: CSG Grunsell, M-E
Raw, FWG Hill (eds): *Veterinary Annual*,
29th issue; Wright, London, 59–63

39 Wilesmith JW, Hoinville LJ, Ryan JBM,
Sayers AR (1992) Bovine spongiform encephalopathy: aspects of the clinical picture and analyses of possible changes
1986–1990. *Vet Rec* 130: 197–201

40 Austin AR, Hawkins SAC, Kelay NS, Simmons MM (1994) New observations on
the clinical signs of BSE and scrapie. In:
R Bradley, B Marchant (eds): *Transmissible Spongiform Encephalopathies*. Proceedings of a Consultation on BSE with
the Scientific Veterinary Committee of
the Commission of the European Communities held in Brussels, 14–15 September 1993; Document VI/4131/94-
EN. European Commission, Agriculture,
Brussels, 277–287

41 Austin AR, Simmons MM (1993) Reduced rumination in bovine spongiform
encephalopathy and scrapie. *Vet Rec*
132: 324–325

42 Austin AR, Simmons MM, Wells GAH
(1997) Pathological temperament
changes in bovines. *Ir Vet J* 50: 304–309

43 Mayhew IG (ed) (1989) *Large Animal
Neurology: A Handbook for Veterinary
Clinicians*. Lea & Febiger, Philadelphia/
London

44 Bradley R, Matthews D (1992) Subacute,
transmissible spongiform encephalopathies: current concepts and future
needs. *Rev Sci Tech Off nt Epiz* 11: 605–
634

45 Scientific Veterinary Committee (1994)
Protocols for the Laboratory Diagnosis
and Confirmation of Bovine Spongiform
Encephalopathy and Scrapie. A report
from the Scientific Veterinary Committee
1994, European Commission, Directorate General for Agriculture, Unit for
Veterinary Legislation and Zootechnics.
Brussels, Belgium

46 Wells GAH, Bradley R, Dawson M, Wilesmith JW (2000) Bovine spongiform encephalopathy. In: *OIE Manual of Stan-*

dards for Diagnostic Tests and Vaccines.
Fourth edition. Office International des
Epizooties, Paris, 457–466

47 Fraser H, Dickinson AG (1968) The sequential development of the brain lesions of scrapie in three strains of mice.
J Comp Pathol 78: 301–311

48 Fraser H (1976) The pathology of natural
and experimental scrapie. In: R Kimberlin (ed): *Slow Virus Diseases of Animals
and Man*, North-Holland, Amsterdam,
267–305

49 Fraser H, McBride PA (1985) Parallels
and contrasts between scrapie and de-
mentia of the Alzheimer type and age-
ing: strategies and problems for experiments involving lifespan studies. In: J
Traber, WH Gispen (eds): *Senile Demen-
tia of the Alzheimer Type*. Springer-Verlag, Berlin, 250–268

50 Lax AJ, Millson GC, Manning EJ (1983)
Can Scrapie Titres be Calculated Accu-
rately from Incubation Periods? *J Gen
Virol* 64: 971–973

51 Kärber G (1931) Beitrag zur kollektiven
Behandlung pharmakologischer Reihen-
versuche. *Arch Exp Path and Pharm*
162: 480–483

52 Dougherty RM (1964) Animal Virus Ti-
tration Techniques. In: RJC Harris (ed):
Techniques in Experimental Virology.
Academic Press, New York, 169–223

53 Taylor DM, McConnell I, Ferguson CE
(2000) Closely similar values obtained
when the ME7 strain of scrapie agent
was titrated in parallel by two individuals in separate laboratories using
two sublines of C57BL mice. *J Virol
Methods* 86: 35–40

54 Kimberlin RH, Walker CA (1978) Pathogenesis of mouse scrapie: effect of route
of inoculation on infectivity titres and
dose response curves. *J Comp Pathol*
88: 39–47

55 Kimberlin RH (1994) A scientific evaluation of research into bovine spongiform
encephalopathy (BSE). In: R Bradley, B
Marchant (eds): *Transmissible Spongiform Encephalopathies*. A Consultation
on BSE with the Scientific Veterinary
Committee of the Commission of the
European Communities held in Brussels,

14–15 September 1993. Document VI/4131/94-EN. European Commission, Agriculture, Brussels, 455–477

56 Kimberlin RH (1996) Bovine spongiform encephalopathy and public health: some problems and solutions in assessing the risk. In: L Court, B Dodet (eds): *3rd International Symposium on Transmissible Subacute Spongiform Encephalopathies: Prion Diseases*, 18–20 March 1996, Paris. Elsevier, Amsterdam, 487–502

57 Fraser H, Bruce ME, Chree A et al. (1992) Transmission of bovine spongiform encephalopathy and scrapie to mice. *J Gen Virol* 73: 1891–1897

58 Masel J, Jansen VAA (2001) The measured level of prion infectivity varies in a predictable way according to the aggregation state of the infectious agent. *Biochim Biophys Acta* 1535: 164–173

59 McLean AR, Bostock CJ (2000) Scrapie infections initiated at varying doses: an analysis of 117 titration experiments. *Philos Trans R Soc Lond (Biol)* 355 (1400): 1043–1050

60 Robinson MM, Cheever DB, Gorham JR (1990) Organ-specific modification of the dose-response relationship of scrapie infectivity. *J Infect Dis* 161: 783–786

61 Baron T (2002) Mouse models of prion disease transmission. *Trends Mol Med* 8: 495–500

62 Telling GC (2000) Prion protein genes and prion diseases: studies in transgenic mice. *Neuropathol Appl Neurobiol* 26: 209–220

63 Flechsig E, Manson JC, Barron R et al. (2004) Knockouts, knockins, transgenics, and transplants in prion research. In: SB Prusiner (ed): *Prion Biology and Diseases* Second Edition. Cold Spring Harbor Laboratory Press, New York, 373–434

64 Scott M, Peretz D, Ridley RM et al. (2004) Transgenetic investigations of the species barrier and prion strains. In: SB Prusiner (ed): *Prion Biology and Diseases* Second Edition. Cold Spring Harbor Laboratory Press, New York, 435–482

65 Aguzzi A, Brandner S, Fischer MB et al. (2001) Spongiform encephalopathies: insights from transgenic models. In: MJ Buchmeier, IL Campbell (eds): *Advances in Virus Research, Vol 56*, Academic Press Inc. San Diego, 313–352

66 Bosque PJ, Ryou C, Telling G et al. (2002) Prions in skeletal muscle. *Proc Natl Acad Sci USA* 99: 3812–3817

6 Cell Culture Models of TSEs

Sylvian Lehmann, Jerome Solassol and Veronique Perrier

Contents

1   Introduction

During the past two decades, considerable efforts have been made to set up tissue culture models of genetic and infectious forms of TSEs and cellular cultures supporting TSE agent replication [1]. These models present, from both a fundamental and applied point of view, several advantages in comparison with *in vivo* models. They provide the ability to analyze at both the molecular and cellular levels, the biological properties of PrP^{C} and PrP^{Sc}. They

Methods and Tools in Biosciences and Medicine
Techniques in Prion Research, ed. by S. Lehmann and J. Grassi
© 2004 Birkhäuser Verlag Basel/Switzerland

also permit to investigate the nature of the infectious agent and the factors governing its propagation and, importantly, they can be used to screen drugs with potential therapeutic value.

In this Chapter, we provide a detailed protocol for the generation of scrapie infected N2a cells [2]. We also give step-by-step procedures to test the biochemical properties (mainly protease resistance and insolubility) of abnormal PrP molecules. These methods are adapted for the detection in cell culture of both PrPSc and mutated PrP (PrPMut) carrying mutations similar to those of individual affected by genetic forms of TSEs [3].

2   Materials

Cell cultures

N2a or Chinese Hamster Ovary (CHO) cell lines can be obtained from the ATCC (reference: CCL-131 and CCL-61, respectively). They are handled and cultured in standard ways. It is wise to split the cells before they reach complete confluence, to record the number of passages and to make frozen stocks of the cells.

Solutions, reagents and buffers

Most reagents are standard (mainly purchased from Sigma). However, references of specific reagents used in our protocols are listed below.
- Proteinase K (PK): 25 mg, ref: 161 519, Roche Diagnostic. Stock at 2 mg/ml in ddH$_2$O, make small aliquots conserved at -20 °C.
- Pefabloc: 100 mg, ref: 1 429 868, Roche Diagnostic. Stock at 100 mM in ddH$_2$O, make small aliquots conserved at -20 °C.
- OptiMEM: ref 51985042, Invitrogen.
- Phosphatidylinositol-specific phospholipase C (PI-PLC): ref: P8804, 5 UI/ml, Sigma.
- Anti-PrP antibodies: see the Chapter related to antibodies in this book.

Buffers:
- Lysis buffer #1 (LB#1), Composition: 150 mM NaCl, 0.5% Triton X-100, 0.5% sodium deoxycholate and 50 mM Tris-HCl (pH 7.5). Store at 4 °C.
- Lysis buffer #2 (LB#2), Composition, 100 mM NaCl, 0.5% NP-40, 0.5% sodium deoxycholate and 10 mM Tris-HCl (pH 8). Store at 4 °C.

Materials

In addition to standard laboratory equipment, the following centrifuges were used in the protocols:
- Refrigerated eppendorf centrifuge reaching 20,000 g
- Ultratop centrifuge (Beckman TLA-100) and dedicated eppendorf able to resist high g force.

3   Methods and protocols

3.1   Generation of chronically prion infected cells

We describe here the generation of prion infected N2a cells. A similar protocol can be used for other cell lines like the GT-1 cells [2]. Our general procedure is to leave the cells in contact with infected mice brain homogenates.

Part A: Preparation of brain homogenates
Brain homogenates are prepared from mice (CD1 or C57Bl/6) in the terminal phase of prion infection with the Chandler/RML or the 22L strains. A 10% (10 g/ 100 ml) brain homogenate in sterile NaCl 9‰ – glucose 5% buffer (or PBS – glucose 5%) is realized either by dounce homogenization or by serial passages through syringes. This homogenate is frozen at –80 °C in small (0.5 ml) aliquots to avoid repeated freeze/thaw. Presence of PrP^{Sc} in the aliquots is confirmed at the time of the cell infection (see Chapter 8). Homogenates from matched unaffected mice are also required as controls.

Dilutions of brain homogenates to a final concentration of 2% or below are made in OptiMEM. To avoid contaminations by conventional agents, we advise "decontamination" of the homogenates with one of these two alternative methods:

1. Warm the 10% homogenate for 20 min at 80 °C in a heat-block and sonicate the sample for 3 min to break the clumps resulting from the heating.
2. After dilution of the homogenate in the OptiMEM, filter the homogenate through a 0.22 µm sterile filter. This filtration can also be done following the previous procedure.

In our tests, pretreatment of the homogenate reduced by one log the efficiency of the cell culture infection [4]. However, it was also noticed that this procedure significantly reduced the toxicity of the homogenate for the cultured cells.

Part B: Infection of the cells
Cells were grown in 6-well plates and plated at 2×10^5 cells/well two days before infection. On the day of infection, the cell are expected to be 70–80% confluent. To infect the cells, different dilutions of a 10% brain homogenate (see above) can be used. We suggest starting with two dilutions: 2% and 0.2% of infectious and control brain homogenate.

- Step 1: Rinse each well with warm sterile OptiMEM for 3 min, remove the OptiMEM and add 1 ml per well of the dilution of the brain homogenate in OptiMEM. Incubate for 5 h in the incubator, then add 1 ml of complete medium and incubate the plate overnight (14–16 h) at 37 °C.
- Step 2: Rinse each well separately twice with phosphate buffered saline (PBS), add 2 ml of fresh medium and incubate for an additional 24 h.

- Step 3: Split each well into a new well of a 6-well plate, as well as in one or two additional 25 cm² flasks. At confluence, lyse one 25 cm² flask to perform a proteinase K (PK) resistance assay (see below). The other flask could serve as a backup or its cells can be frozen for further subcloning to isolate infected clones.
- Step 4: Passage the cells up to 10 times and monitor the presence of PrP^{Sc}. In the first or the second passage after infection, the PrP^{Sc} from the inoculum could still be detected.

Troubleshooting
Toxicity and contamination: depending on the origin of the homogenate, the cultured cells may not survive the infection or be contaminated by conventional agents. To minimize this problem, we suggest using a lower homogenate concentration or pretreating the homogenate by heating and/or filtration as indicated above. Subcloning of the infected cells is sometimes needed and could be performed at the first passage after the inoculation to maximize the chance of getting infected clones.

3.2   Cell-to-cell transmission of prion infection

To confirm the generation of infectivity by the newly infected cells, the best way is to inoculate cell lysates in mice. However, an alternative way is represented by the possibility to use cell cultures to detect the presence of infection as follows:
- Step 1: Grow control and infected cells to obtain confluent 75 cm² flasks. Rinse the cultures twice in PBS, scrape the cells in 5 ml of cold PBS under sterile conditions and collect them in a conical 15 ml falcon tube.
- Step 2: Pellet the cells by spinning the tubes for 6 min at 800 xg. Remove the supernatant and resuspend the pellet in 100 µl of cold sterile PBS with 5% glucose. To prepare the cell extract, submit this suspension to four cycles of freezing-thawing in liquid nitrogen. The inoculum is then passed through a 27-gauge needle several times. 50 µl of the preparation is diluted in 1 ml of Opti-MEM and used to infect cells following the procedure 3.1. above.

3.3   Detection of proteinase K resistant PrP in cell culture

Proteinase K (PK) digestion of PrP is the most common method of detecting PrP^{Sc} and the presence of prions. PrP^{mut} is less resistant to PK digestion than PrP^{Sc} and a lower concentration of the enzyme has to be employed to distinguish it from wild-type PrP.

Protocol 1   PrP^{Sc} detection

This protocol describes the detection of PrP^{Sc} in scrapie infected N2a or GT1 cell [2].

1. Start with a subconfluent tissue culture flask of 25 cm^2 or equivalent. Gently rinse the flask twice for 2 min with cold PBS.
2. Aspirate the PBS, put in the flask 0.6 ml of lysis buffer 1 (LB 1) eventually completed with protease inhibitors (pepstatin and leupeptin, 1 µg/ml and EDTA, 2 mM). Rock the flask for 10 s and lay it flat at 4 °C for 15 min. The protease inhibitor phenylmethyl sulfonyl fluoride (PMSF) is a PK inhibitor and must be avoided.
3. Collect the cell lysate in a 1.5 ml eppendorf, spin for 3 min at 10,000 xg in a microcentrifuge, a procedure that removes debris/DNA but does not pellet significant amounts of PrP. Put the supernatant in a new eppendorf tube, keep on ice.
4. Measure the total protein concentration with the BCA assay (Pierce). Adjust the protein concentrations of the samples to 0.6 mg/ml by adding LB 1 to the tubes.
5. Take 50 µl of each lysate and mix it in a new eppendorf with an equal volume of Laemmli sample buffer with dithiothreitol (DTT) (LMSB) 2X, boil for 5 min at 90 °C. These tubes will be used to evaluate PrP^C levels in the samples.
6. To detect PrP^{Sc}, put 500 µl of the samples (the equivalent of 300 µg of protein) in eppendorfs containing 2.4 µl of the stock solution of PK at 2 mg/ml (ratio: PK/protein 1/62, equivalent to 16 µg of PK/mg protein). Gently vortex and put at 37 °C for 30 min. Put the tubes on ice, add 5 µl of Pefabloc in each (stock at 100 mM: 1 mM final), mix, wait 5 min.
7. Centrifuge the eppendorfs at 14,000 rpm (20,000 xg) for 45 min at 4 °C, gently discard the supernatant by returning the tubes and resuspend the pellet first in 30 µl of Luria Betani (LB), then by adding 30 µl of LMSB 2X, vortex, boil for 5 min at 90 °C.
8. Perform the detection of PrP by Western blot using a 12% SDS PAGE. Always load a positive control (e.g., PK digested prion infected brain homogenate). To detect PrP^{Sc}, an anti-PrP antibody raised against the carboxy-terminus of the protein must be used. In fact, in prion infected cells, most of PrP^{Sc} is cleaved around codon 90 and is not recognized by antibodies raised against the amino-terminus including those raised against the octapeptide repeats. *Note:* a detailed Western blot protocol can be found in Chapter 8.

Protocol 2   PrP^{Sc} detection

The main difference with the protocol described above is the use of a slightly different lysis buffer and the absence of a short centrifugation after the lysis of the cells. The duration of the enzymatic digestion is also 60 min instead of 30 min and the ratio between PK and proteins can vary from 1/25 (1 µg of PK for 25 µg of proteins) to 1/50 (1 µg of PK for 50 µg of proteins) which allows a variation in the intensity of the PrP^{Sc} signal.

1. Aspirate the culture medium from a subconfluent tissue culture flask of 25 cm^2 and rinse the cells twice with 1 volume of cold PBS.

2. Add 0.4 ml of cold lysis buffer 2 (LB 2) and leave the flask for a few minutes, at room temperature.

3. Collect the cell lysate in a 1.5 ml eppendorf, and remove the white material (nuclei+DNA) with a pipette.

4. Measure the total protein concentration with the BCA assay (Pierce). Adjust the protein concentrations to 1 mg/ml with LB2.

5. Remove 20 µl of lysate from each tube and mix it in a new eppendorf tube with an equal volume of Laemmli sample buffer (LMSB) 2X, boil for 5 min at 90 °C. These tubes will be used to evaluate total PrP levels in the samples.

6. Perform the PK digestion on 400 µl of adjusted lysate by adding 10 µl of PK diluted at 0.8 mg/ml (ratio of 1/50 which corresponds to 1 µg of PK for 50 µg of proteins). Gently vortex, then incubate the samples at 37 °C for 60 min. When the enzymatic digestion is done, add 5 µl of Pefabloc (stock at 100 mM: 1 mM final), mix, and incubate for 5 min on ice.

7. Centrifuge the tubes at 14,000 rpm (20,000 xg) for 30 min at 4 °C, then gently discard the supernatant by returning the tubes. Resuspend the pellet first in 20 µl of LB, and then add 20 µl of LMSB 2X, boil the samples for 5 min at 90 °C. Samples can be stored for severals weeks at –20 °C before doing the Western blot.

8. Perform the detection of PrP by western blot (see Protocol 1, step 8).

Protocol 3 Rapid protocol for PrP^{Sc} detection

This protocol is an alternative to the previous protocols. It permits a rapid detection of PrP^{Sc} in small culture wells and can be useful to screen therapeutic agents. However, this protocol which does not include a measure of the total protein concentration or a centrifugation step does not permit a precise quantitation of PrP^{Sc}.

1. Start with confluent wells of a 12-well plate. Gently rinse the wells once with cold PBS. Put 1 ml of cold PBS in the wells, scrape the cells and collect them in 1.5 ml eppendorfs.
2. Pellet the cells by spinning the tubes for 3 min at 5,000 xg in a microcentrifuge. Remove the supernatant; add 50 µl of LB 2, vortex to put the pellet in suspension in the buffer, put on ice for 10 min.
3. Vortex the tube again and pellet the debris by spinning the tubes for 3 min at 10,000 xg in a microcentrifuge.
4. Put 30 µl of the supernatant in a new eppendorf containing 5 µl of a solution of PK at 0.2 mg/ml. Gently vortex and put at 37 °C for 30 min.
5. Add to each tube 30 µl of LMSB 2X, vortex, boil for 5 min at 90 °C.
6. Perform the detection of PrP by Western blot (see Protocol 1, step 8).

Protocol 4 Standard protocol for PrP^{Mut}

This protocol describes the detection of PrP^{Mut} in CHO transfected cells [3]. Steps 1 to 5 are the same as in Protocol 1 for PrP^{Sc} detection.

6. To detect PrP^{Mut}, put 500 µl of the samples (the equivalent of 300 µg of protein) in an 2 ml eppendorf containing 5 µl of a solution of PK at 0.2 mg/ml (ratio: PK/protein 1/300, equivalent to 3.3 µg of PK/mg protein). Gently vortex and put at 37 °C for 30 min. Put the tube on ice, add 5 µl of Pefabloc (stock at 100 mM: 1 mM final), mix, wait 5 min.
7. Methanol precipitate the samples as follows: fill the eppendorf with methanol stored at –20 °C, mix by inversion several times, put at –20 °C for at least 2 h. Centrifuge the eppendorfs at 14,000 rpm (20,000 xg) for 10 min at 4 °C, remove the supernatant by inversion, let the pellet dry (but not too much!), resuspend the pellet first in 30 µl of LB 1, then by adding 30 µl of LMSB 2X, vortex, boil for 5 min at 90 °C.
8. Perform the detection of PrP by Western blot (see Protocol 1, step 8).

Troubleshooting

Importantly, PK digestion must be optimized for each cell line in order to completely digest PrP^C from control, non-infected cells.

3.4   Detection of insoluble PrP in cell culture

Insolubility in non-ionic detergent is a common property of PrP[Sc] (in scrapie infected GT1 cells for example) but also of PrP[Mut] when expressed in CHO or in N2a cells [3, 5].

Protocol 5   Detection of insoluble PrP in cell culture

Steps 1 to 4 are the same as in Protocol 1 for PrP[Sc] detection.

  5. To detect insoluble PrP, put 400 µl of the samples in Beckmann eppendorfs and centrifuge at 70,000 rpm for 30 min in the TLA 100.4 rotor of a Beckman Optima TL ultracentrifuge to separate detergent-soluble and detergent-insoluble protein. An alternate method valid for PrP[Sc] in N2a cells consists of spinning the samples in a regular refrigerated centrifuge at 14,000 rpm (20,000 xg) for 45 min at 4 °C.
  6. At the end of the centrifugation, take 200 µl of the supernatant, put in new eppendorfs and methanol precipitate (see Protocol 4, step 7).
  7. Carefully remove the supernatant left in the eppendorfs after the spinning of step 5. The pellet often looks like a small oil drop on the bottom side of the tube. Resuspend this pellet in 40 µl of LMSB 1X, vortex, boil for 5 min at 90 °C.
  8. Perform a western blot detection of soluble and insoluble PrP on 12% SDS PAGE: load equal volume of each samples and Western blot with anti-PrP antibodies (either against the amino-terminus to detect full length PrP or against the carboxy-terminus to detect cleaved PrP). Load a positive control (brain from scrapie infected animals) on the gels and always spin control lysate containing wild type PrP[C] to test the efficacy of the separation. *Note:* a detailed Western blot protocol can be found in Chapter 8.

3.5   Triton X-114 partition of PrP (hydrophobicity property)

Triton X-114 partition before or after PIPLC digestion is a common way to test the presence of a GPI anchor attached to a protein. PrP[Sc] or PrP[Mut] expressed in cultured cells have their GPI anchor not completely cleaved (in these non-denaturing conditions) and display a property that we have called "hydrophobicity" or "PIPLC resistance" [3].

Protocol 6   Realisation of a Triton X-114 partition

The following protocol describes the realisation of a Triton X-114 partition as done for PrP^{Mut} in CHO transfected cells.

1. Starts from a sub-confluent Petri 35 mm dish or equivalent. Gently rinse the cells once with cold PBS. Put 1 ml of cold PBS in the dish, scrape the cells and collect them in an eppendorf.

2. Pellet the cells by spinning the tubes for 3 min at 5,000 xg in a microcentrifuge. Remove the supernatant, add 0.55 ml of 1% Triton X-114 1 (diluted in cold PBS from a 12% stock, see [6] for preparation) complete with protease inhibitors (pepstatin and leupeptin, 1 µg/ml and EDTA, 2 mM). Vortex, leave on ice for 20 min.

3. Centrifuge the eppendorf at 14,000 rpm (20,000 xg) for 1 min, put the supernatant in a new eppendorf.

4. Put the tube at 37 °C for 10 min. Perform the first phase separation by spinning the tube at 14,000 rpm (20,000 xg) for 1 min, collect the aqueous phase (A1) at the top of the tube, put it in a new eppendorf and methanol precipitate (see Protocol 4, step 7).

5. Add to the detergent phase 0.5 ml of cold PBS complete with protease inhibitors, put on ice for 5 minutes, vortex to obtain a homogenous mix. Prepare two new eppendorfs per sample, one labelled "PIPLC" and containing 5 µl of PI-PLC. Split the samples in 2 aliquots of 0.25 ml in the eppendorfs, vortex, and leave on ice for 2 h.

6. Put the tube at 37 °C for 10 min. Perform the second phase separation by spinning the tube at 14,000 rpm (20,000 xg) for 1 min, collect the aqueous phase (A2) at the top of the tube, put it in the new eppendorf and methanol precipitate both the A2 fraction and the detergent phase (D) that remained in the tube (see Protocol 4, step 7).

7. Western blot an equal volume of the aqueous (A2) and detergent phases of control and PIPLC treated samples. A regular glycosyl-phosphatidyl-inositol (GPI) anchored protein is present in the detergent phase before PIPLC treatment and shifts to the aqueous phase thereafter. PrP^{Mut} are retained significantly more in the detergent phase. *Note:* a detailed Western blot protocol can be found in Chapter 8.

3.6   Other methods

Other methods relevant to prion infected cells are listed below:
- Subcloning of cells to isolate prion susceptible cell lines: see [7]
- Filter retention assay: see [8]
- Scrapie cell assays: see [9]
- Steel wire assays: see [10]
- Screening of drugs: see for example: [11–17]

Acknowledgements

Our group is supported by the CNRS ("Centre National de la Recherche Scientifique") and by grants from the GIS prion ("Groupement d'intérêt scientifique" sur les Prions) and the European Community (Biotech, QTRL-2000-02353).

References

1   Béranger F, Mangé A, Solassol J et al. (2001) Cell culture models of prion diseases. *Biochem Biophys Res Com* 289: 311–316

2   Nishida N, Harris DA, Vilette D et al. (2000) Successful transmission of three mouse-adapted scrapie strains to murine neuroblastoma cell lines overexpressing wild-type mouse prion protein. *J Virol* 74: 320–325

3   Lehmann S, Harris DA (1996) Mutant and infectious prion proteins display common biochemical properties in cultured cells. *J Biol Chem* 271: 1633–1637

4   Lehmann S, Laude H, Harris DA et al. (2001) *Ex vivo* transmission of mouse adapted prion strains to n2a and gt1–7 cell lines. In: K Iqbal, SS Sisodia, B Winblad (eds): *Alzheimer's Disease: Advances in Etiology, Pathogenesis and Therapeutics*. John Wiley & Sons, Ltd., Chichester, UK, 680–686

5   Priola SA, Chesebro B (1998) Abnormal properties of prion protein with insertional mutations in different cell types. *J Biol Chem* 273: 11980–11985

6   Bordier C (1981) Phase separation of integral membrane proteins in triton x-114 solution. *J Biol Chem* 256: 1604–1607

7   Bosque PJ, Prusiner SB (2000) Cultured cell sublines highly susceptible to prion infection. *J Virol* 74: 4377–4386

8   Winklhofer KF, Hartl FU, Tatzelt J (2001) A sensitive filter retention assay for the detection of prp(sc) and the screening of anti-prion compounds. *FEBS Lett* 503: 41–45

9   Klohn PC, Stoltze L, Flechsig E et al. (2003) A quantitative, highly sensitive cell-based infectivity assay for mouse scrapie prions. *Proc Natl Acad Sci USA* 100: 11666–11671

10 Flechsig E, Hegyi I, Enari M et al. (2001) Transmission of scrapie by steel-surface-bound prions. *Mol Med* 7: 679–684

11 Caughey B, Raymond GJ (1993) Sulfated polyanion inhibition of scrapie-associated prp accumulation in cultured cells. *J Virol* 67: 643–650

12 Mange A, Nishida N, Milhavet O et al. (2000) Amphotericin b inhibits the generation of the scrapie isoform of the prion protein in infected cultures. *J Virol* 74: 3135–3140

13 Kocisko DA, Baron GS, Rubenstein R et al. (2003) New inhibitors of scrapie-associated prion protein formation in a library of 2000 drugs and natural products. *J Virol* 77: 10288–10294

14 Korth C, May BC, Cohen FE, Prusiner SB (2001) Acridine and phenothiazine derivatives as pharmacotherapeutics for prion disease. *Proc Natl Acad Sci USA* 98: 9836–9841

15 Rudyk H, Vasiljevic S, Hennion RM et al. (2000) Screening congo red and its analogues for their ability to prevent the formation of prp-res in scrapie-infected cells. *J Gen Virol* 81: 1155–1164

16 Supattapone S, Nguyen HO, Cohen FE et al. (1999) Elimination of prions by branched polyamines and implications for therapeutics. *Proc Natl Acad Sci USA* 96: 14529–14534

17 Perrier V, Wallace AC, Kaneko K et al. (2000) Mimicking dominant negative inhibition of prion replication through structure-based drug design. *Proc Natl Acad Sci USA* 97: 6073–6078

7   PrP^{Sc} Immunohistochemistry

Olivier Andréoletti

Contents

1   Introduction

Historically, diagnosis of transmissible spongiform encephalopathies (TSEs) has been a matter for pathologists. For decades, the only available diagnostic tool in this field was the identification of spongiform changes and gliosis in CNS sections, and conventional histopathology still remains the method of choice for the confirmation of clinically suspected TSE.

In the early 1980s, the discovery of scrapie-associated fibrils (SAFs) [1] and the production of antibodies against SAFs [2] were the first steps of a revolution in TSE investigation. SAFs are protein aggregates specifically obtained from infected tissue homogenates in the presence of detergents. Later on, the main constituent

Methods and Tools in Biosciences and Medicine
Techniques in Prion Research, ed. by S. Lehmann and J. Grassi
© 2004 Birkhäuser Verlag Basel/Switzerland

of SAFs was identified as the prion protein, PrP, which is a host-encoded protein. Numerous cell types express PrP in its normal isoform (PrP^C). In incubating animals, PrP accumulates as an abnormal isoform, PrP^{Sc}, in various tissues long before the occurrence of spongiform changes in the brain or the appearance of clinical signs. PrP^{Sc} exhibits amylogenic properties and is distinguishable from PrP^C by its greater resistance to proteolysis and high temperatures [3].

In recent years, monoclonal antibodies (mAbs) have been produced that specifically recognise PrP^{Sc} as found in SAFs from a TSE-infected brain [4, 5]. However, so far, none of these mAbs has proved suitable for direct identification of PrP^{Sc} in immunohistochemistry. Most anti-PrP antibodies used for that purpose are unable to distinguish between the protein isoforms: they usually bind PrP^C but bind PrP^{Sc} poorly unless it is subjected to a denaturation process. For instance, in amyloid deposition of PrP^{Sc} in tissues most PrP epitopes are buried. Therefore, a specific and sensitive PrP^{Sc} *in situ* detection requires both the suppression of PrP^C immunoreactivity and a denaturation step (PrP^{Sc} antigen retrieval) to enhance the immunoreactivity of the amyloid [6]. The most efficient pretreatments used to enhance the specific immunolabelling of PrP^{Sc} consist of an incubation of the tissue sections in formic acid [7] followed by a hydrated autoclaving step at 121 °C [8]. It is worth noting that formic acid treatment is also an efficient means of inactivating TSE infectivity in fixed tissues (see Chapter 11 in this book). Such chemical and heat treatment requires the use of formalin-fixed, paraffin-embedded tissue sections.

The choice of appropriate anti-PrP antibodies for immunohistochemistry is crucial for good quality labelling. Both polyclonal and monoclonal antibodies can be used for this purpose. The last decade has seen an amazing proliferation of anti-PrP antibodies. It is noteworthy that antibodies suitable for immunohisto-chemistry are not necessarily adapted to Western blot or ELISA. This is very likely due to the pretreatment of the slides and the aldehyde fixation of tissues, which generates particular conformational alterations of the protein. Another criterion in choosing an antibody is the species to be studied. Some antibodies raised against highly conserved parts of the protein are very reactive in one species and poorly reactive in others. The precise behaviour of an antibody is difficult to predict and each should therefore be tested to establish its particular reactivity.

Furthermore, PrP^C expression levels vary among tissues. The highest PrP^C expression is usually observed in the central nervous system (CNS), and is 50- to 200-fold higher than in the lymphoid tissues [9]. This may be associated with specificity problems for PrP^{Sc} labelling in the brain and, as a result, some antibodies are suitable for the investigation of lymphoid tissues and difficult to use for the CNS.

Several workshops have aimed to draw up consensus immunohistochemistry protocols for TSE diagnosis in humans [10] and animals (EU FAIR project – PL98–7021). However, immunohistochemists can be compared to cooks: most of them prefer their own recipe. No standard protocol has therefore emerged, and different protocols are still described in the literature.

This Chapter will present some robust techniques for preparation of tissue samples and immunohistochemical detection of PrP^{Sc}. Special sections will deal with more complex situations like double labelling, cell culture detection and tissues from transgenic mice expressing high PrP^{C} levels.

2 Materials

Materials
- Paraplaste Plus – Sherwood Laboratories-LABONORD, ref 06948430
- Superfrost plus slides, CML, ref LLR3PLUS
- Cytoblock (50 reactions), Shandon, ref 74010150
- Dako Pen, Dako, ref S2002
- Flow cytometry 5 ml tubes, Becton Dickinson
- LaBtek cell culture chamber slides, Becton Dickinson, ref 354108

Chemicals
- Alkaline phosphatase-streptavidin complex, Dako, ref P0397
- Bovine serum albumin, Sigma, ref A8327
- EnVision™ anti-mouse peroxidase, Dako, ref K 4001or K 4003
- Faramount mounting medium, Dako, ref S3025
- Formic acid 98%, Merck, ref 405792
- Goat anti-mouse IgG – biotin labelled, Dako, ref E0431
- Goat anti-rabbit Ig – biotin-labelled, Dako, ref E0432
- Goat anti-mouse Fab2′ Ig – FITC labelled, Dako, ref F0479
- Goat normal serum, Dako, ref X0907
- Histozol, Shandon, ref HISTOSOL+
- Guanidium isothiocyanate, Sigma, ref 50985
- Permanent mounting medium, Dako, ref S3026
- Proteinase K, Roche, ref 1.373.200
- Mayer's haematoxylin, CML, ref 320554
- Sheep normal serum, Dako, ref X0503
- Streptavidin-peroxidase complex, Dako, ref P0397
- Tris buffered saline (TBS) 10 × pH 7.65 , Microm, ref 30 PA-0235
- 3-amino-9-ethylcarbazole (AEC), Dako, ref K3461
- 3,3′-diaminobenzidine (DAB) liquid, Dako, ref K3466
- 5-bromo-4-chloro-3-indolyl phosphate/nitroblue tetrazolium (BCIP/NBT), Dako, ref K0598

Solutions

Solution 1: Buffered formalin 10%:
- Formalin 40%   2.5 L
- Distilled water   22.5 L
- Monobasic sodium phosphate (NaH_2PO_4/H_2O)   100 g
- Anhydrous dibasic sodium phosphate (NaH_2PO_4)   162.5 g
- NaOH 4 N   100 ml
- Magnetic stir-up agitation and heating (30 °C) are needed to dissolve the different components efficiently. Store in the dark.

Solution 2: Sodium citrate/Citric acid solution:
- Solution A: Citric acid 0.1 M
  Monohydrated citric acid (Sigma, Paris, France, ref S4641) 21.01 g
  Distilled water   to make 1,000 ml
- Solution B: Sodium citrate 0.1 M
  Sodium citrate (Sigma, Paris, ref C1909)   29.41 g
  Distilled water   to make 1,000 ml
- Working solution should be prepared extemporaneously by adding 18 ml of solution A to 82 ml of solution B and 900 ml of distilled water. The pH is adjusted to 6.1–6.2 with 1 M NaOH.

Solution 3: Endogenous peroxidase inhibitor:
- Methyl alcohol   200 ml
- Hydrogen peroxide 30%   2 ml
- Solution 3 should be prepared extemporaneously.

Solution 4: Tris buffered saline pH 7.65 (TBS):
- 10X Tris (0.5 M) buffered saline (8.9–9%)
  Tris base   60.57 g
  Sodium chloride   89–90 g
  Distilled water   to make 1,000 ml
- Test and adjust pH to 7.65 using hydrochloric acid (1 M)
- Store stock 10X solution in refrigerator
- Working solution (0.05 M TBS) should be prepared extemporaneously by diluting stock solution 1/10 in distilled water.

Solution 5: TBS washing solution:
- Tris buffered saline pH 7.65   1 L
- Skimmed lyophilised milk   10 g
- Tween 20   500 μl

3   Methods

3.1   General method

PrP^{Sc} immunohistochemistry of tissue sections is done in three steps: (1) proper tissue handling and slide preparation (fixation and paraffin embedding), (2) antigen retrieval ensuring the destruction of PrP^{C} and unmasking of PrP^{Sc}, (3) immunoenzymatic detection.

Protocol 1   Tissue and slide preparation

Tissue fixation
1. Fixation with a buffered formalin solution (solution 1) will enhance the quality of fixation and will avoid the formation of precipitate in tissues. Tissues are fixed in 10% buffered formalin solution for 4–10 days before being cut into 2-mm thick slices, put into cassette and fixed again for 24 h.
2. The cassette or tissue can be immersed for 1 h in 98% formic acid under a fume hood to reduce infectivity titre. This has no major effect on immuno-histochemistry quality. This procedure is very interesting when handling mouse brain samples for lesion profiling. It renders the tissues harder and easier to section.
3. Periods longer than two months for tissue fixation will affect the quality and intensity of PrP^{Sc} labelling and make the tissue more difficult to process. To avoid over-fixation, the biopsy samples (like tonsil biopsies or brain biopsies) should not be fixed more than 24–48 hours before paraffin embedding. If the dehydration is not possible within this period, the samples can then be rinsed for one hour in tap water and thereafter stored in 80° alcohol for up to several weeks.

Tissue dehydration
Tissue samples are generally dehydrated and paraffin impregnated using an automated device. However, manual procedures, with mixing, remain possible. The quality of tissue dehydration will greatly influence the resistance of slides to the pretreatment protocol. Dehydration quality depends on (1) the quality of solvents (renewed when needed), (2) the quantity of tissue/total volume of solvent available, and (3) respect of a minimum length dehydration protocol. The vacuum inclusion automated device will reduce the length of the protocol. As an example, we describe a dehydration programme which can be used for medium-size tissue samples in an automated atmospheric pressure device.
1. Formalin (solution 1) 1 h,
2. Tap water 30 min
3. 80° alcohol 90 min
4. 95° alcohol 1 h (bath 1)
5. 95° alcohol 1 h (bath 2)

   6. Absolute alcohol 90 min (bath 1)
   7. Absolute alcohol 90 min (bath 2)
   8. Toluene or histozol 90 min (bath 1)
   9. Toluene or histozol 90 min (bath 2)
 10. Paraffin (Paraplaste plus) 90 min
 11. Paraffin 90 min

Tap water should be renewed at each cycle, while 80° alcohol and the first 95° alcohol, absolute alcohol and toluene baths should be renewed every 5–7 cycles. The second 95° alcohol, absolute alcohol and toluene baths should be renewed only every 10–12 cycles (depending on the nature of the treated tissues).

Slide preparation
Tissue sections, 2–5 µm thick, should be collected on adhesive-treated slides (Superfrost plus, for example) and dried overnight at 56 °C before further processing. The use of high-quality slides (electrostatically treated) will allow the sections to resist pretreatment.
Slides are then deparaffinised and rehydrated
   1. 10 min toluene (or equivalent like histozol)
   2. 10 min toluene (or equivalent like histozol)
   3. 5 min acetone
   4. 5 min acetone
   5. Rinse 5 min in tap water
Solvent quality is crucial for good quality deparaffinisation. Do not hesitate to renew them often. Toluene used for deparaffinisation can be recycled in a tissue dehydration automated device.

Protocol 2   Antigen retrieval

Basic PrP^{Sc} antigen retrieval requires two steps
   1. Sections are immersed in a 98% formic acid bath for 30 min under a chemical fume hood, before rinsing in tap water. Formic acid breaks the PrP^{Sc} amylogenic structure and enhances the destruction of PrP^{C} immunoreactivity by heat.
   2. Sections are then heated at 121 °C, under pressure (2.1 bar), in 10 mM citrate buffer (pH 6.1–6.2) (solution 2).
Citrate buffer (10 mM, pH 6.1–6.2) efficiently enhances PrP^{Sc} immunoreactivity when compared to distilled water. A pressure cooker, rather than an autoclave, seems to be the best way to carry out heating since the duration of heating is more easily controlled.

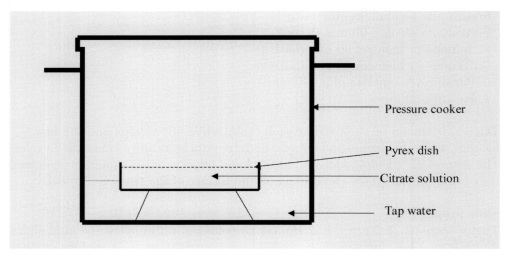

Figure 1   Schematic representation of a slide heating "water bath" in pressure cooker

Sections should not be immersed directly into the bottom of pressure cooker but should be placed in a water bath system that allows uniform diffusion of heat and better preservation of histological structures (see Fig 1). When water is boiling at the bottom of the cooker, the slides are immersed in citrate solution and the pressure cooker is closed. Make sure that boiling water will not overflow into the plate containing citrate solution. Time is counted from the moment the safety pressure detection system is switched on (pressure over 1.3 bar).

The heating duration may vary with the application, the nature of the tissue and the anti-PrP antibody. Five min is a minimum, but heating can be extended to 30 min. The intensity of PrP^{Sc} labelling is usually correlated with the heating duration, but will reach a plateau generally after 10 or 20 min.

Shorter heating (5–10 min) will dramatically reduce PrP^{C} immunoreactivity, although longer heating could restore it partially. Using high PrP^{C} expressing tissues with high-affinity antibody will require short (10 min) rather than long heating. When using a new antibody, it is advisable to optimise heating duration.

To stop heating, the pressure cooker is placed under a stream of cold tap water before opening. The sections are allowed to cool in citrate solution for 10–20 min before being rinsed in tap water. After this treatment, slides can be left in water for several hours without obvious modification of immunolabelling results.

Protocol 3   PrP^{Sc} immunodetection

All operations are performed at room temperature in a humid chamber.

1. Endogenous peroxidase inhibition: 30 min by immersion in 0.3% (w/w) hydrogen peroxide methyl alcohol bath prepared immediately before use (solution 3). This inhibition of peroxidase activity will last for 12 h.
2. Rinse sections in tap water for 5 min.
3. Use a hydrophobic pen (Dako) to draw a circle on the slide at least 1 mm around the tissue section.
4. Equilibrate slides in TBS solution (solution 4) for 5 min.
5. Nonspecific binding blockade: shake the slides to discard liquid. Incubate sections with 20% normal goat serum in TBS (solution 4, not TBS washing solution) for 20 min. 150–300 µl are required depending on the tissue surface.
6. Discard liquid. Do not rinse.
7. Apply 150–250 µl of the anti-PrP antibody at the chosen dilution for 45–60 min. Primary antibody is diluted in 0.3% BSA TBS solution.
8. Discard liquid. Rinse three times (5 min each) in TBS washing solution (solution 5).
9. Discard liquid and apply 150–250 µl of biotinylated secondary goat anti-rabbit or anti-mouse immunoglobulin, diluted in TBS (solution 4) with 5% normal sheep serum, for 30 min.
10. Discard liquid – Rinse twice in TBS washing solution (solution 5), 5 min each.
11. Discard liquid and apply a streptavidin-peroxidase complex (1:100 diluted in TBS, solution 4) for 30 min.
12. Rinse twice in TBS (solution 4).
13. Apply DAB solution (Dako). The optimal labelling is typically obtained after 5 min, but the reaction should be observed using low-magnification microscope to stop it at the optimal time. Rinse in tap water to stop the reaction.
14. Sections are counter-stained in Mayer's haematoxylin for 2 min.
15. Rinse in tap water for 5 min.
16. Dehydrate by rapid immersion (20 s) in 95° alcohol, absolute alcohol and toluene baths (twice each).
17. Mount in appropriate hydrophobic medium (Dako, Permanent mounting medium).

Controls

Controls are essential for the interpretation of PrP^{Sc} immunolabelling. Each immunohistochemistry run should include at least one previously identified positive control and a positive control in which anti-PrP antibody is omitted. The latter guarantees the absence of secondary antibody nonspecific binding. A control in which the primary antibody is replaced by isotype-matched immunoglobulins (monoclonal antibody) or normal serum (polyclonal antibodies) is necessary to evaluate primary antibody nonspecific binding.

3.2   Transgenic mouse tissues

The recent development of transgenic mice expressing high PrP^C levels in most of their tissues has set a challenge for immunohistochemists. PrP^{Sc}-specific labelling depends on the capacity of the pretreatment protocol to reduce PrP^C antigenicity, but also on the choice of anti-PrP antibody. Antibodies with high affinity for PrP^C should be avoided, as they generate persistent background staining.

Optimal results are typically obtained using:
1) Formic acid for 30 min.
2) Hydrated autoclaving in citrate solution (solution 2) for 5–10 min.
3) Mild proteolysis using a 5 µg/ml proteinase K (PK)/TBS solution (solution 4) at 37 °C for 5–15 min. The length of digestion will vary according to the level of PrP^C expression depending on the mouse line. A cold tap water bath is used to stop the PK activity.

The procedure should be performed with this sequence since heating sections after mild proteolysis would unmask undigested PrP^C. Proteolysis can be extremely deleterious for tissue structures and the quality of slide preparation is important. In most cases, PK reduces specific PrP^{Sc} labelling, and the duration of proteolysis should be optimised in each case.

3.3   Double labelling

Double labelling is a highly informative but delicate technique. In prion diseases, it allows phenotypic identification of PrP^{Sc}-positive cells [11]. The granular appearance of PrP^{Sc} deposits makes double labelling easier than in other applications. One of the main limits of double labelling in PrP^{Sc} applications is the requirement for cell phenotype markers suitable for paraffin-embedded tissue.

Different strategies can be chosen for double labelling, but one rule should be borne in mind: the greater the number of steps, the higher the background. The strategy proposed here is a two-step method based on the simultaneous application of the two primary antibodies, each raised in different species, followed by a simultaneous incubation with two secondary antibodies, each specific for the species of origin of the primary antibody, and each carrying a separate enzyme label. Most commercially available antibodies for cell phenotyping are monoclonal mouse antibodies. Joint use with a good anti-PrP rabbit polyclonal serum may be helpful.

The protocol steps are the same as described in single labelling (protocol 3) except for steps 7, 11 and 13–16.

Protocol 4   PrP^{Sc} double labelling

7. Primary antibody reaction

Both primary antibodies are applied as a single mix. Typically, pairs of primary antibodies consist of a mouse monoclonal antibody (cell phenotype marker) and a rabbit polyclonal antibody (anti-PrP serum). Each antibody included in the mix must be used at its optimal dilution as determined by single-labelling protocols. The antibodies are diluted in TBS (solution 4) containing 0.3% BSA.

9. Secondary antibody incubation

Optimal labelling requires signal amplification for both the anti-PrP and the cellular phenotype. Both secondary antibodies are mixed and applied together. To reduce cross-reactivity, secondary antibodies have to be obtained from the same species (in our example goat).

The first secondary antibody, chosen to reveal the cell phenotype marker, is a goat antibody specific for mouse Ig and directly coupled to a dextran polymer carrying peroxidase (EnVision™). The other secondary antibody revealing the anti-PrP antibody is a classical biotin-labelled goat antibody, specific for rabbit (1:200 diluted). The mixed secondary antibodies are applied for 30 min at room temperature. This mix has to be supplemented with 5% normal sheep serum to reduce non-specific tissue binding.

11. PrP^{Sc} signal amplification

An alkaline phosphatase-streptavidin complex (1:100 diluted) is applied for 30 min at room temperature to amplify the PrP^{Sc}-specific signal.

13–16: Staining of enzymatic activity and slide mounting

Staining of peroxidase and phosphatase enzymatic activities is performed sequentially. To optimise staining, these steps should be followed in order under the microscope at low magnification.

PrP^{Sc} is stained first. Alkaline phosphatase substrate, BCIP/NBT (Dako-black deposits), is applied for 5–10 min. After 10 min, non-specific background staining increases rapidly. After removal of the liquid, rinse the slides in TBS washing solution (solution 5) to stop the reaction. Peroxidase activity is visualized using AEC (Dako-red deposits) for 3–20 min. The sections are then rinsed in tap water before Mayer's haematoxylin counterstaining. The slides must be mounted in aqueous medium (Faramount mounting medium) since NBT/BCIP and AEC fade in organic solvent.

Controls

For double labelling, cross-reactivity controls are paramount. They should be performed for each pair of primary antibodies and each sample, in order to check the absence of inter-species reactivity of the secondary antibodies toward the primary antibodies. Possible interaction between the two secondary antibodies must be discounted. Controls are obtained by applying each primary antibody alone together with both secondary antibodies.

3.4 Application to cell culture

Infected cell lines are useful for the investigation of the prion pathogenesis at the cellular level. Different methods are available for *in situ* PrP^C and PrP^{Sc} detection from cell cultures.

Protocol 5   PrP^C/PrP^{Sc} direct labelling in fixed cells

This immunofluorescence technique is easy to perform and very informative. However, it does not allow a discriminatory labelling of PrP^{Sc}, since PrP^C immunoreactivity has not been eliminated. Nevertheless, the presence of PrP^{Sc} is characterised by granular labelling, while the labelling of PrP^C will look more diffuse.

The labelling is done as follows:

1. Place the cells on culture chamber slides (LaBtek) at low density – wait for 2–4 days. The cells should remain separated. Do not allow confluence.
2. Wash twice with cold TBS (solution 4, 4 °C).
3. Fix with 10% formalin in isotonic NaCl for 20 min at room temperature (RT).
4. Treat with 1% NH$_4$Cl in TBS for 5 min at RT. This step reduces background staining due to formalin fixation.
5. Rinse twice in TBS (solution 4).
6. Permeabilise the cells with 0.5% Triton X-100 in TBS (solution 4) for 5 min.
7. Rinse twice in TBS (solution 4).
8. Treat with 3 M guanidium isothiocyanate for 5 min.
9. Rinse 5 × 2 min in TBS.
10. Equilibrate the slides in TBS washing buffer (solution 5) for 5 min.
11. Blocking non-specific binding sites: Discard the liquid, incubate the sections with 20% normal goat serum in TBS (solution 4, not TBS washing solution) for 20 min. Discard the liquid – do not rinse.
12. Apply the primary antibody at the previously determined dilution. The antibody is diluted in TBS containing 0.1% BSA. The incubation can be performed for 1 h at RT or overnight at 4 °C.
13. Rinse twice, 5 min each, in TBS washing buffer (solution 5).
14. Apply FITC-labelled goat anti-mouse or rabbit (depending on the anti-PrP antibody) secondary antibody diluted 1/15 in TBS with 5% normal sheep serum, 45 min at RT.
15. Rinse for 5 min in TBS washing buffer (solution 5).
16. Rinse in TBS for 5 min to eliminate Tween.
17. Mount the slides in TBS/glycerol (vol/vol) before microscope examination. This mounting system allows short-term storage (24–96 hours) at –20 °C.

Protocol 6   Cell culture specific PrP^{Sc} labelling

In this method, cells are fixed and paraffin-embedded prior to classical PrP^{Sc} immunohistochemistry. This procedure is delicate and requires some skills. It can be carried out using 200,000 cells but is optimal with one to three million cells, and a cell-block preparation system like Cytoblock™ from Shandon.

1. Scrape the cells from the culture flasks.
2. Dissociate by pipetting.
3. Pellet the cells by centrifugation at 1,200 rpm for 3 min in flow cytometry 5 ml tubes.
4. Wash in 4 ml isotonic NaCl and pellet by centrifugation at 1,800 g for 3 min.
5. Resuspend the cells by pipetting in 1 ml of 2% formalin diluted in isotonic NaCl solution (initially 40%). DO NOT use buffered formalin solution (solution 1). Allow to stand for 3 min.
6. Add 4 ml isotonic saline solution before pelleting the cells by centrifugation.
7. Wash twice in isotonic NaCl solution by successive resuspension/centrifugation of the pellet.
8. Use the cytoblock procedure as described by the manufacturer for the inclusion of cells in paraffin.

Perform the antigen retrieval and immunolabelling procedures as described in the single labelling section. For use of transgenic cells (with high PrP^{C} expression level), mild PK proteolysis (2–5 min) can be used as described in the transgenic mice section.

4   Troubleshooting

4.1   Tissue lost from slides

When performing such drastic pretreatments required for specific PrP^{Sc} labelling, it is very common that the sections may loosen if they do not adhere strongly to the slides. The reasons are multiple and somewhat difficult to identify. Some critical points that require particular attention have already been described. Check the dehydration/paraffin embedding procedure, as well as the quality of the slides and the paraffin. A recently renewed formic acid solution may also generate this problem, particularly with tissues like spleen, which are difficult to handle at sectioning.

4.2 PrP^C residual labelling in tissue

This problem is common when using transgenic mice that overexpress PrP^C, but may also be observed in particular situations (particularly primary antibodies) with CNS tissue from conventional animals. In such cases, controls, in which secondary antibodies have been omitted, should be used. Background staining is generally dusty and more intense in particular neuro-anatomical structures (grey matter) and cells (neurons).

Several solutions can be explored:
1) Reduce the duration of the heating step during pretreatment.
2) Apply a longer PK digestion to pretreated slides, but note that overly long PK digestion will also reduce PrP^{Sc}-specific labelling.
3) Reduce the concentration of the primary antibody.

Primary antibody may be the cause of such background staining. Some antibodies have such a high affinity for PrP^C that it is impossible to eliminate background staining using them.

4.3 Background staining due to nonspecific primary antibody binding

In this situation, controls are needed in which the primary antibody has been omitted. Background staining is more or less uniform and affects connective tissue and blood vessels.

These problems can be overcome by:
1) Diluting the primary antibody and performing a conventional immunoreaction.
2) Diluting the primary antibody (about 1/100) and performing an overnight reaction at 4 °C.
3) Increasing the concentration of BSA in the primary antibody diluent.

Despite those ameliorations, some primary antibodies, and more particularly rabbit immune sera, may still generate nonspecific background staining. Such primary antibodies should be avoided.

5   Applications

In situ detection of PrP^Sc constitutes one of the main methods in TSE investigations. It is especially useful in the case of inconclusive histopathology and offers a firm basis for diagnosis, using paraffin-embedded material. In addition, the main uses of this method are (i) localisation of PrP^Sc in tissues, and (ii) identification of cell populations involved in deposition of PrP^Sc.

Numerous studies on experimental and natural TSE pathogenesis have been carried out with a PrP^Sc immunohistochemical approach in order to determine the phenotype of the cells involved in the infectious processes and to determine the dissemination pathways within the host [12–14].

For strain typing using mouse inoculation, PrP^Sc distribution in the brain can be used as a complementary tool in the characterization of TSE strains, even if the official method is based on the distribution and intensity of spongiotic changes.

In isolated cells or permanently infected cell lines, Western Blotting is traditionally used to identify PrP^Sc. *In situ* detection of PrP^Sc can also be useful for evaluating the proportion of infected cells and the distribution of PrP^Sc in subcellular compartments [15].

Acknowledgements

Many thanks to S. Benestad from the National Veterinary Institute of Norway (NVI) for her critical reading of the manuscript.

References

1   Merz PA, Somerville RA, Wisniewski HM et al. (1981) Abnormal fibrils from scrapie-infected brain. *Acta Neuropathol (Berl)* 54: 63–74

2   Diringer H, Rahn HC, Bode L (1984) Antibodies to protein of scrapie-associated fibrils. *Lancet* 2: 345

3   Harris DA (1999) Cellular biology of prion diseases. *Clin Microbiol Rev* 12: 429–444

4   Paramithiotis E, Pinard M, Lawton T et al. (2003) A prion protein epitope selective for the pathologically misfolded conformation. *Nat Med* 9: 893–899

5   Zou WQ, Zheng J, Gray DM et al. (2004) Antibody to DNA detects scrapie but not normal prion protein. *Proc Natl Acad Sci USA* 101: 1380–1385

6   Van Everbroeck B, Pals P, Martin JJ et al. (1999) Antigen Retrieval in Prion Protein Immunohistochemistry. *J Histochem Cytochem* 47: 1465–1470

7   Kitamoto T, Ogomori K, Tateishi J et al. (1987) Formic acid pretreatment enhances immunostaining of cerebral and

systemic amyloids. *Lab Invest* 57: 230–236

8   Haritani M, Spencer YI, Wells GA (1994) Hydrated autoclave pretreatment enhancement of prion protein immunoreactivity in formalin-fixed bovine spongiform encephalopathy- affected brain. *Acta Neuropathol* 87: 86–90

9   Moudjou M, Frobert Y, Grassi J et al. (2001) Cellular prion protein status in sheep: tissue-specific biochemical signatures. *J Gen Virol* 82: 2017–2024

10 Bell JE, Gentleman SM, Ironside JW et al. (1997) Prion protein immunocytochemistry UK five centre consensus report. *Neuropathol Appl Neurobiol* 23: 26–35

11 Andreoletti O, Berthon P, Levavasseur E et al. (2002) Phenotyping of protein-prion (PrP[Sc])-accumulating cells in lymphoid and neural tissues of naturally scrapie-affected sheep by double-labeling immunohistochemistry. *J Histochem Cytochem* 50: 1357–1370

12 Van Keulen LJ, Schreuder BE, Meloen RH et al. (1996) Immunohistochemical detection of prion protein in lymphoid tissues of sheep with natural scrapie. *J Clin Microbiol* 34: 1228–1231

13 Sigurdson CJ, Spraker TR, Miller MW et al. (2001) PrP(CWD) in the myenteric plexus, vagosympathetic trunk and endocrine glands of deer with chronic wasting disease. *J Gen Virol* 82: 2327–2334

14 Foster JD, Parnham DW, Hunter N et al. (2001) Distribution of the prion protein in sheep terminally affected with BSE following experimental oral transmission. *J Gen Virol* 82: 2319–2326

15 Vilette D, Andreoletti O, Archer F et al. (2001) *Ex vivo* propagation of infectious sheep scrapie agent in heterologous epithelial cells expressing ovine prion protein. *Proc Natl Acad Sci USA* 98: 4055–4059

Western Immunoblotting Techniques for the Study of Transmissible Spongiform Encephalopathies

Michael J. Stack

Contents

1   Introduction

Initial diagnosis of transmissible spongiform encephalopathies (TSEs) has always been made by the detection of neuronal vacuolation within formaldehyde-fixed brain sections by histopathological examination. Over time, specific antibodies for pathological markers of TSEs have been produced and immunology-based techniques are being increasingly used to aid diagnosis for screening purposes and to provide information at the molecular level for the disease group as a whole. Western immunoblotting is an established technology and, due to its

Methods and Tools in Biosciences and Medicine. Techniques in Prion Research, ed. by S. Lehmann and J. Grassi. Birkhäuser Verlag Basel/Switzerland. © Crown copyright 2004. Published with the permission of the Controller of Her Britannic Majesty's Stationary Office. The views expressed are those of the author and do not necessarily reflect those of Her Britannic Majesty's Stationary Office or the VLA or any other government department.

use of specific antibodies, it is a tool that is used across many scientific disciplines to provide diagnosis of particular diseases, or answers to research problems. This chapter outlines some of the Western immunoblotting techniques which are being used to study TSEs, discusses the advantages and disadvantages of each technique, and makes suggestions on the circumstances under which each may be best used.

One of the disease characteristics of TSE is the presence of an abnormal protease enzyme-resistant protein, initially termed the *prion* protein (PrP) [1] or the *scrapie* protein, PrP^{Sc}, [2, 3]. Prion protein or PrP^{Sc} is thought to be derived from a normal *cellular* host protein, termed PrP^c, which is particularly abundant within central nervous system tissue. In an unaffected animal PrP^c with a molecular weight of 33–35 kDa (PrP33–35) is completely digested after protease enzyme treatment. However, in an affected animal, the PrP^c is partially resistant to the enzyme. Under conventional conditions the protease treatment leads to the removal of 62 N-terminal amino acids , leaving a core fragment of some 141 amino acids, often referred to as PrP^{res} with a molecular weight of 27–30 kDa (PrP27–30) [3].

In the case of TSE's the quality of results obtained from the application of Western immunoblotting techniques is reliant on using an efficient method of extracting PrP^c and PrP^{res} from central nervous system (CNS) tissue. The extraction techniques currently in use have all developed from painstaking initial research targeted at isolation of the TSE agent. Research to track down the agent responsible started at least as far back as 1964 when it was shown that 99% of the detectable scrapie activity of brain was associated with particulate elements rich in membranes [4]. This led to the membrane hypothesis, as it was shown that it was not possible to separate the scrapie agent from membranous cellular elements [5]. This took the concept of agent purification, by subcellular fractionation of brain tissue, a stage further by using a collection of elaborate differential centrifugation and sucrose density gradient techniques to isolate particular cell components, and checking the relative scrapie activity in mice. Development of purification protocols were further advanced when the effects of nuclease and protease digestions and sedimentation characteristics were studied, and gel electrophoresis and chromatography techniques were introduced [6, 7].

In the early 1980's the characterisation and detection of PrP was greatly enhanced by the identification of scrapie associated fibrils (SAFs) or prion rods. Initially described as long abnormal fibrils, the SAFs were described in extracts of infected scrapie and Creutzfeldt-Jakob disease (CJD) brain tissues using extractions from crude synaptosomal-mitochondrial brain fractions treated with octyl-glucoside and subjected to discontinuous sucrose density gradient centrifugation [8]. These fractions contained the highest number of infectious units for CJD. For spleen extractions the same method was used, but spleens were homogenised in 0.2 M KCl in the first instance. Progress in purification of PrP^{Sc} from hamster brains led to the identification of PrP27–30 and of amyloid birefringent rods, termed prion rods that were composed of this protein [9]. This lengthy purification protocol included TritonX-100/sodium deoxycholate extraction, polyethylene glycol precipitation, nuclease and proteinase digestion, cholate/sodium lauryl sarcosine extraction, $(NH_4)_2SO_4$ precipitation, and further Triton X-100 sodium dodecyl sulphate (SDS) extraction and sedimentation through discontinuous sucrose gradients followed by centrifugation using a vertical rotor. This method was scaled up by a factor of five and zonal rotor centrifugations were added. Fractions from the gradients were traced using radiolabelling with N-succinimidyl 3-(4-hydroxy-[5–^{125}] iodophenyl) proprionate. The use of SDS extraction and radiolabelling to check purity in this protocol enabled the identification of prion rods as PrP with the 27–30 kDa molecular weight.

Later, antibodies produced against PrP27–30 which were immunogold labelled, were used in conjunction with electron microscopy to demonstrate that prion rods were composed of this protein. The N-terminal amino acid sequence of these infectious rods was identical to that determined for PrP27–30 [10]. The present Western immunoblotting techniques for PrP detection described in this Chapter owe much to the fortitude of previous researchers and the forerunner of most purification techniques used today was originally described as a rapid and efficient method to enrich SAF-protein from hamster scrapie brains [11]. The method was a combination of sarcosyl detergent extraction, ultracentrifugation and proteinase K digestion.

Using sarcosyl and proteinase K based Western immunoblotting techniques, the PrP^c is destroyed in uninfected tissue during the detergent extraction and enzymic steps, but any remaining PrP^{res} from infected tissue can be detected using an antibody raised to a particular amino acid sequence of the PrP protein; a sequence not destroyed by protease treatment. In general, PrP^{res} gives three characteristic molecular weight bands which correspond to recognition by the antibody of the diglycosylated, monoglycosylated and unglycosylated regions of the protein. These bands are detected by the initial application of polyacrylamide gel electrophoresis [12] to separate the three bands by molecular weight, followed by the use of Western immunoblotting and interaction with the appropriate antibody for visualisation purposes [13]. This technique has the advantage of being able to detect PrP^c or PrP^{res} by the omission or addition of a protease enzyme within the method. As well as the comparison of molecular weight values, the blots obtained for PrP^{res} can be compared for glycoform ratio [14, 15]. To obtain this ratio, the combined magnitude of signals of all three protein bands is defined as 100% and the contribution of each band calculated as a percentage of the whole. To obtain a comparison of ratio between samples, the percentage signal from the diglycosylated band is plotted against the percentage signal obtained for the monoglycosylated band and shown as positions on a scattergraph.

2   Western immunoblotting methods

2.1   Office International des Epizooties (OIE) method

The first Western immunoblotting technique to be used for diagnostic purposes is presented in Protocol 1 and used a relatively large amount of starting material (4 g) and lengthy ultracentrifugation steps in order to concentrate any PrP^{res}. The technique was developed from earlier published methods [16, 17]. Although it has become known colloquially as the Office International des Epizooties (OIE) technique the earliest record of its publication was within a report published by the European Commission, Directorate General for Agriculture, Unit for Veterinary Legislation and Zootechnics, Brussels [18] and was subsequently published in a graphic format within a book Chapter [19]. Evidence from comparative experiments within a European Union Commissioned project (FAIR PL 987021) suggest that methods that have larger amounts of starting material and ultracentrifuge steps are more sensitive than some of the more rapid tests. In some field cases where inconclusive results have been obtained with the usual statutory tests, and enough brain tissue was available, the OIE technique or similar techniques have been able to give a positive result; the most recent example being the report of cases of scrapie with unusual features in Norway [20]. This technique may also be a useful confirmatory test when histopathological examination and immunohistochemistry results are unobtainable or inconclusive due to autolysis of the tissue e. g. in fallen stock surveys. As the extraction method used in this technique also produces SAFs, which are detectable by negative stain electron microscopy [19], it is capable of supplying two confirmatory results, a SAF result and an immunoblotting result. Samples can be examined for SAFs before the denaturation stage for subsequent western immunoblotting. Disadvantages of the technique include; the amount of starting material required, (optimum weight of 4 g), the length of the protocol, and the need for ultracentrifugation equipment. These disadvantages exclude its use as a mass screening tool where a quick turn a round time for a test is required, such as for abattoir samples.

Reagents
- Solution 1 – Brain Lysis Buffer (BLB)
  10 g N-Lauroylsarcosine, sodium salt (SIGMA L5125) in 100 ml of 0.01 M sodium phosphate buffer pH 7.4 prepared fresh for each sample preparation procedure.
  Preparation of 0.01 M sodium phosphate buffer:
  a) dissolve 0.345 g NaH_2PO_4 in 250 ml distilled water
  b) dissolve 0.9 g Na_2HPO_4 in 500 ml distilled water
  c) adjust to pH 7.4 by adding the acidic NaH_2PO_4 solution to the basic Na_2HPO_4 solution-store at 4 °C

- Solution 2–1 M Tris/HCl pH 7.4
- Solution 3–100 mM PMSF: 0.435 g Phenylmethylsulfonylfluoride in 25 ml propan-1-ol stored in a dark bottle at 4 °C
- Solution 4–100 mM NEM: 0.313 g N-ethyl-maleimide in 25 ml propan-1-ol store in a dark bottle at 4 °C
- Solution 5 – Proteinase K (PK): 1 mg/ml (in distilled water)
- Solution 6–20% sucrose: 20 g sucrose in 100 ml of 10% KI HSB
- Solution 7 – KI-HSB: 1.5 g Sodium thiosulphate, 1.0 g N-lauroyl-sarcosine in 1 ml 1 M Tris/HCl add 10 g (for 10%) or 15 g (for 15% solution) potassium iodide make up to 100 ml with distilled water – store at 4 °C
- Solution 8–1 × sample buffer   2 ml SDS (20%)
  1 ml Tris/HCl (1 M, pH 7,4)
  1 ml Mercaptoethanol
  0.6 g sucrose
  1–3 drops Bromphenolblue
  15 ml distilled water

Protocol 1   OIE technique incorporating ultracentrifugation

The starting amount of brainstem material (preferentially obex region for cattle) to be analysed may depend on origin/quality of the sample and amount of material available but in general 4 g of tissue is used.

1. Take appropriate amount of brainstem material (obex region).
2. Cut into small pieces (carefully remove the dura mater) and add 5 ml of BLB (Solution 1) with 10 µl of 100 mM PMSF (Solution 3) and 10 µl of 100 mM NEM (Solution 4).
3. Homogenise in a glass homogenisator (douncer).
4. Put it into a 50 ml vessel.
5. Add another 5 ml of BLB (Solution 1) into the douncer and rehomogenise if necessary, and combine with first homogenate. Sonicate for 1 min.
6. Centrifuge at 20,000 g (17,000 rpm) for 30 min at 10 °C in a 70 Ti Beckmann ultracentrifuge rotor.
7. Carefully tip the supernatant into clean centrifuge tubes and centrifuge at 177,000 g (46,000 rpm) for 2 h 30 min at 10 °C in a 70 Ti Beckmann ultracentrifuge rotor.
8. Discard the supernatant and suspend the pellet in 3 ml distilled water with 50 µl 1 M Tris/HCl pH 7.4 (0.0167 M) (Solution 2) by gentle aspiration with a pipette.
9. Incubate in a water bath at 37 °C for 15 min whilst stirring.
10. Add 6 ml of 15% KI-HSB (Solution 7) and incubate for a further 30 min as in step 9.
11. Divide the solution into two aliquots of 4.5 ml (only if samples are divided into aliquots for plus and minus PK treatment).
12. To one aliquot add 1 mg/ml PK (Solution 5) and incubate for 1 h as in step 9 amount of PK solution to be added depending on weight of starting material:
    1 g + PK: 22.5 µl
    2 g + PK: 45 µl
    4 g + PK: 90 µl
13. To the other aliquot add 4.5 ml of 10% KI-HSB (Solution 7) and then overlay the total onto 2 ml of 20% sucrose (Solution 6) and centrifuge at 189,000 g (51,000 rpm) for 1 h at 10 °C in a 70 Ti Beckmann ultracentrifuge rotor.
14. Carry out step 13 on the PK treated sample.
15. Carefully tip off supernatants.
16. Resuspend samples in 40 µl 1 × sample buffer (Solution 8).
17. Heat for 5 min at 95 °C prior to loading.
18. Follow steps 3 to 9 of Protocols 4 and 5 for electrophoresis and immuno-blotting.

2.2 Enrichment of PrP^{res} using sodium chloride precipitation

Another western immunoblotting method has also recently been described in which the extraction process enriches for PrP^{res} using a sodium chloride precipitation. The technique was considered to be well suited for TSE strain and species type analyses and may also be of use for purification prior to mass spectrometry studies, given the lack of complicated reagents used in the processing method. Furthermore, this technique was reported to obtain an end product containing up to 97.4% of the total PrP^{Sc} in the starting tissue [21].

Reagents
- Solution 1 – Homogenisation Buffer: 0.5% IGEPAL CA-630 (NP-40) and 0.5% sodium deoxycholate in phosphate buffered saline (PBS) pH 7.4.
- Solution 2 – PMSF: 5 mM PMSF (Sigma-Aldrich), protease inhibitor.
- Solution 3 – PBS: Phosphate Buffered Saline (PBS) pH 7.4.
- Solution 4 – NaCl Sarc: 20% Sodium Chloride in PBS containing 0.1% sarcosyl.
- Solution 5 – Tris Sarc: 25 mM Tris-HCl buffer, pH 6.8, containing 0.05% sarcosyl (w/v).
- Solution 6 – Loading Buffer – 10% v/v glycerol, 50 mM Tris pH 6.8, 2% w/v SDS, 3% β-mercaptoethanol.

Protocol 2 Enrichment of PrP^{res} using sodium chloride precipitation

1. 10% homogenate of CNS tissue using a Polytron homogeniser (Kinematica, Switzerland) in cold homogenisation buffer (Solution 1). Polytron at setting 4 twice for 6 s each. An OmniGLH homogeniser (CAMLAB, UK) works equally well.
2. 100 μl of homogenate is diluted with an equal volume of homogenisation buffer (Solution 1) and incubated with proteinase K (Sigma-Aldrich) for 1 h at 37 °C with mild rocking. (Proteinase K used at 30 μg per ml of homogenate for mouse scrapie, BSE and sheep scrapie and 50 μg for sporadic CJD samples). The reaction is stopped by the addition of a protease inhibitor, PMSF (Sigma-Aldrich) up to a final concentration of 5 mM (Solution 2).
3. Add 300 μl of PBS (Solution 3) and make up to a 10% NaCl concentration by the addition of an equal volume of NaCl Sarc (Solution 4). The tube is kept on ice for 10 min with occasional shaking.
4. Centrifugation at 16,000 g for 10 min at room temperature and the pellet washed twice in Tris Sarc (Solution 5) and then centrifuged again (16,000 g × 10 min)
5. The resultant pellet is resuspended in 2.5 × loading buffer (Solution 6) and loaded onto gels. For a starting volume of 100 μl of homogenate (10 mg brain equivalent) the final pellet is resuspended in the same volume (100 μl) of the loading buffer (Solution 5), of which 30 μl (3 mg brain equivalent) is loaded per lane for the Western blot analysis.
6. Samples are subjected to SDS PAGE and transferred to a PVDF membrane as in other protocols and incubated at room temperature with the primary antibody for 1 h.
7. The membrane is washed with PBS (Solution 3), and incubated again for 35 min with the corresponding secondary antibody. The primary antibodies used in this protocol are SAL1 (manuscript in preparation – Theodoros Sklaviadis) or the commercially available monoclonal 6H4 (Prionics AG, Switzerland).
8. Following three washes of 15 min each the reaction was visualised using the CDP star (New England Biolabs) chemiluminiscence detection technique according to manufacturer instructions.
9. The membrane is exposed to X-OMAT (Kodak) X-ray films for between 5 s to 5 min depending on the degree of positivity in the starting material.
10. Films are scanned (Hewlett Packard Scanjet5 6300C) and the positive signal analysed and expressed as integrated optical density using Gel-Pro Analyser 3.0 software.

2.3 Sodium phosphotungstic acid precipitation technique

The Spongiform Encephalopathy Advisory Committee (SEAC) in the UK has put forward recommendations for the surveillance of scrapie in sheep and these include the postmortem testing of animals being slaughtered at abattoirs. This type of surveillance would help in assessing the number of sheep that may be incubating scrapie that are passing through abattoirs. In the absence of a live animal test to assess preclinical prevalence the testing of non-neural tissues has assumed considerable importance, as any move away from taking brain tissue would give greater practical flexibility in an abattoir situation. However, differences in tissue densities and the smaller amounts of PrP^{res} present in non-neural tissues demand modifications to the technique.

The main mode of primary infection for some of the TSEs is by way of oral ingestion and the alimentary tract, [22] (scrapie), [23] (bovine spongiformence-phalophathy – BSE), [24] (kuru), [25] (transmissible mink encephalopathy), but the sequence of events as to the tissue entry site, multiplication sites and transportation mechanism of the agent into the CNS remains obscure. Infectiv-ity of many ovine non-neural tissues has been demonstrated by bioassay in mice [22]. There is increasing evidence of PrP^{res} detection by immunhistochemistry and western immunoblotting techniques. PrP^{res} has been detected in tissues from ovine spleen and lymph nodes [26, 27, 28] More recently immunohisto-chemical staining techniques of ovine lymphoid tissues from 55 clinically affected sheep have shown PrP^{res} to be present in 54 (98%) of these sheep; in the spleen, retropharyngeal lymph node, mesenteric lymph node and the palatine tonsil. PrP^{res} was also detected in tracheobronchial (93%), prefemoral (87%) and prescapular (88%) lymph nodes as well as in the solitary lymphoid follicles or Peyer's patches of the ileum in 24 (89%) of the 27 sheep in which lymphoid tissue was present in the sections of the ileum [29] The immunos-tained protein has also been reported in tonsils of experimental scrapie-infected sheep long before the occurrence of clinical signs [30] and there has also been the report of the detection of abnormal PrP staining within tonsillar germinal centres by immunohistochemistry in a tonsil biopsy, from a human patient believed to have died due to the new variant CJD (vCJD) [31]. Western immunoblot analysis of this human tonsil material revealed the presence of a protease resistant protein with the sizes and the intensity ratios of the three PrP immunoblotting bands being similar to those seen in the immunoblot analysis of the brain from the same patient.

In the case of peripheral routes of infection in experimental rodent scrapie models invasion of the CNS is thought to be preceded by agent replication in lymphoid tissue such as spleen, lymph nodes and the Peyer's patches of the intestine. Infection is thought to spread from visceral sites of replication along the splanchnic nerves, to the mid-thoracic spinal cord region, to the rest of the CNS and to other peripheral nerves [17, 32–35]. In scrapie-affected mice, PrP^{res} can be detected in non-neural tissues months before the clinical disease is seen

[17, 36]. Others have reported finding PrP[res] in lymphoid tissues of sheep in the preclinical and clinical stages of scrapie infection by Western immunoblotting [37–39]. In this Chapter we describe a developed technique that has been successfully used on a selection of non-neural sheep tissues, spleen, mesenteric lymph nodes, distal ileum and tonsils from clinically suspect scrapie cases (Protocol 3). The method gives a 100% sensitivity when compared to the IHC results for the same tissues (unpublished results from the author) and is based on the incorporation of a sodium phosphotungstic acid (NaPTA) precipitation step to enhance the PrP[res] band signals. The NaPTA technique originated from a published report on the use of a conformation dependent immunoassay (CDI) to compare different scrapie strains in hamsters [40]. The method was then developed and modified into a Western immunoblotting procedure to study the tissue distribution of PrP[res] in vCJD cases [41] The NaPTA technique described in this Chapter is a further modification.

Reagents

- Solution 1 – DPBS: Dulbecco's PBS solution: Dissolve one vial of Dulbecco's Phosphate Buffered Saline (D-7030 Sigma Aldrich) in 900 ml of H_2O, make up to 1 litre with H_2O and pH to 7.4.
- Solution 2 – SLB: Spleen Lysis Buffer: Dissolve 3 g of Tris (271193G BDH Laboratory Supplies) in 500 ml of H_2O and adjust pH to 7.4. Then add 0.5 g of Magnesium-Chloride 6-Hydrate (101494 V BDH Laboratory Supplies), 2.5 g of N-lauroylsarcosine 2.9 g of Sodium Chloride (102414J BDH Laboratory Supplies) and add 10 ml of Tergitol (NP-40) (Sigma Aldrich).
- Solution 3 – DPBS 4% Sarc: Dulbecco's Phosphate Buffered Saline (PBS) with 4% sarcosyl, dissolve 4 g of N-lauroylsarcosine (L-5125 Sigma Aldrich) in 90 ml of Dulbecco's PBS solution, make up to 100 ml with DPBS (Solution 1).
- Solution 4 – Benzonase:
  Dilute 10 μl of (50 unit/ml) Benzonase (E-1014 Sigma Aldrich) in 4,512 μl of DPBS (Solution 1).
- Solution 5 – NaPTA: Sodium Phosphotungstic Acid, dissolve 1.72 g of Magnesium-Chloride 6-Hydrate (101494 V BDH Laboratory Supplies) and 2 g of Sodium phosphotungstate dibasic hydrate (P-6395 Sigma Aldrich) in 40 ml of H_2O, make up to 50 ml with H_2O. Adjust pH to 7.4 with NaOH.
- Solution 6 – DPBS 0.1% Sarc: Dulbecco's Phosphate Buffered Saline with 0.1% sarcosyl. Dissolve 0.1 g of N-lauroylsarcosine i.n 90 ml of DPBS (Solution 1), make up to 1 litre with DPBS (Solution 1).
- Solution 7 – EAPBS: 250 mM Edetic Acid. Dissolve 4.65 g of Ethylene-diamine-tetra-acetic Acid (E-5134 Sigma Aldrich) in 50 ml of DPBS (Solution 1) and pH to 8.0.
- Solution 8 – Prionics-Check sample buffer.
- Solution 9 – 11.5 mM Magnesium Chloride.
  Dissolve 1.165 g MgCI6 Hydrate (101494v BDH Laboratory Supplies) in 500 ml DPS.
- Solution 10 – Proteinase K prepare 1 mg/ml stock (Roche).

Protocol 3   Sodium phosphotungstic acid precipitation technique for non-neural tissues

Flow diagram of protocol 3 for non-neural tissues

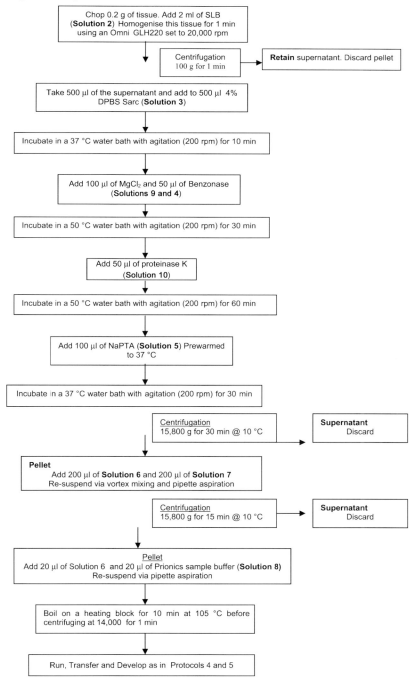

2.4 Prionics-Check technique and VLA hybrid differential Western immunoblotting technique

In addition to its usefulness in diagnostic terms, recent research has concentrated on using Western immunoblotting to discriminate between different TSE strains, or PrP^{res} conformers. For instance, the similarities in the molecular weights and glycoform ratios obtained from Western immunoblotting techniques have supported the likelihood of a bovine origin for vCJD human infections. Molecular weights and glycoform ratios of PrP^{res} derived from sporadic or iatrogenic forms of CJD have been shown to be quite different from those of vCJD and BSE [14, 15]. Measurement of the differences in the molecular weight of the unglycosylated protein band and glycoform ratio analysis have also been shown in rodent scrapie models to try and distinguish strain differences [42–45]. The comparative analysis of long term proteinase K sensitivity of PrP^{res} of murine and hamster strains has also shown discrimination. [40, 42]

Several scientific publications have reported discrimination, to varying degrees, between natural sheep scrapie and sheep experimentally infected with the BSE strain by molecular analysis [46–53]. One of these discriminatory techniques has evolved by modification of a commercially available western immunoblotting technique, Prionics-Check, which is a validated test kit for the diagnosis of BSE [54, 55] and is obtainable from Prionics Ag, Zurich, Switzerland. In a recent study, this modified version of Prionics-Check, which has been designated 'the VLA hybrid technique', was used to compare the molecular weights and glycoform ratio differences between natural sheep scrapie, experimentally inoculated BSE in Romney and Cheviot breed sheep, and natural BSE in cattle [52]. The sheep passaged scrapie strains CH1641 and SSPB1 were also compared as they are often used in Western immunoblotting comparisons due to their characteristic blot patterns, whereby CH1641 gives a similar, but not identical, profile to BSE in sheep and SSBP1 gives a molecular profile akin to natural scrapie samples. CH1641 has been propagated in sheep and was originally isolated from a British Cheviot sheep in 1971 [56] and has been characterised by serial passage in sheep as either a single strain or an unresolved mixture of strains. It has unusual changes in incubation properties on second and third passage in comparison to Group A strains and has been classified as a C Group strain [57, 58]. SSBP1 (Sheep Scrapie Brain Pool 1) has been passaged mostly through Cheviot sheep [59] and is now known to be a mixture of scrapie strains designated as A Group strains [60].

Using the VLA hybrid technique, brain tissue samples were tested in parallel using two monoclonal antibodies (mAbs) raised to specific amino acid sequences of the bovine and ovine PrP. The two antibodies used were mAb 6H4, (Prionics Ag), which is a mouse IgG1 antibody that recognises the sequence in the bovine PrP protein at amino acid positions 144–152 [61], and mAb P4 (BioPharm Ltd) which is also raised in mice, and recognises the amino acid sequence in the ovine PrP protein at amino acid positions 89–104 [62]. The standard Prionics-Check and the modified protocol used for strain typing are amalgamated in Protocols 4 and 5 in this Chapter.

Protocols 4 and 5   Prionics-Check technique and VLA hybrid differential Western immunoblotting technique

The Prionics-Check test is purchased in kit form and all the main reagents are supplied within the kit; a reagent list is therefore not available but by following the Protocol researchers should be able to carry out the methods with little difficulty. The VLA hybrid differential immunoblotting technique is a modified version of Prionics-Check [54], incorporating a centrifugation step [52] and has been used to differentiate between natural scrapie and experimentally challenged BSE in sheep. Omit the centrifugation step for the standard Prionics-Check test.

1. 1.5 ml of a 10% homogenate (Prionics) is centrifuged at 1,127 g for 5 min (TLA 45 rotor – Beckman).
2. Proteinase K (Roche) (10 μg/ml of a 1 mg/ml stock to give a final concentration of 100 μg/ml) was added to 100 μl of the supernatant and this was incubated at 50 °C for 1 h. Pefabloc (Boehringer to 1 mM final concentration?) and sample buffer (100 μl) (Prionics) is added and reacted at 105 °C for 10 min.
3. 10 μl of the Solution is loaded onto 12% Bis-Tris polyacrylamide gels (Invitrogen).
4. Electrophoresis is carried out at 200 v for 35 min and Western immunoblotting on to polyvinylidene difluoride PVDF membrane (Millipore) at 150 v for 1 h.
5. The blots are blocked in 50 ml of blocking buffer (Prionics) for 1 h and incubated overnight at 4 °C in a 1:5000 dilution of primary antibody (6H4 Prionics in blocking buffer).
6. Membranes are washed in Tris Buffered Saline (with 0.05% Tween 20) 4 × 7 min and incubated in the secondary antibody (1:5000) (goat anti-mouse conjugated to alkaline phosphatase) (Prionics) (for 1 h at room temperature).
7. They are then washed again in TBST 4 × 7 min and then incubated in luminescence buffer (Prionics) for 5 min.
8. The labelling is visualised by means of enhanced chemiluminescence system (CPD-Star Tropix). Signals are quantified using Fluor S Multimager computer analysis (Quantity One, software Biorad UK Ltd).
9. Using this system, molecular weights are measured by comparison to the biotinylated markers on the gel and the centre position for each sample band is recorded as the molecular weight.

3   Results and Discussion

All the techniques outlined in this Chapter should ideally give the characteristic blot profile for PrP^res consisting of three distinct protein bands corresponding to the diglycosylated, monoglycosylated and unglycosylated peptides of the PrP molecule as recognised by a specific antibody. In some circumstances (probably to do with epitope presentation due to damage or masking effects) only the diglycosylated band or the diglycosylated and monoglycosylated bands are detected and this is where controls with and without proteinase K treatment, and the molecular weight markers, help in establishing an accurate diagnosis. To be able to successfully carry out strain/species differentiation techniques it is imperative that all three bands are detected and appropriate comparative positive controls are incorporated on each gel to compare molecular weights and glycoform ratio meaurements.

The resultant glycoforms from using the VLA hybrid differential Western immunoblotting technique (Protocol 5) showed that three distinctions could be made between natural ovine scrapie cases and sheep experimentally inoculated with BSE [52]. There were subtle differences in the molecular weight positions of the diglycosylated, mono-glycosylated and unglycosylated forms of PrP^res from the different ruminant TSEs, (Fig. 1 Blot A). In particular, a distinct difference for the unglycosylated protein band was observed. For ovine scrapie samples, this band was noticeably of a higher molecular weight than that found for brain samples from the Romney and Cheviot breed sheep infected with BSE and, to a lesser degree, higher than that observed for bovine BSE samples. The sheep passaged CH1641 scrapie strain gave molecular weights similar to, but not identical to, BSE. The SSBP1 experimental scrapie strain gave molecular weights which were akin to natural scrapie cases. When the antibody mAb P4 was substituted for mAb 6H4 in the technique only the natural scrapie samples and SSBP1 gave strong signals. BSE in sheep and the CH1641 strain gave weak reactions and PrP^res from BSE infected cattle could not be detected at all (Fig. 1 Blot B).

Using the comparison of glycoform ratios, the technique provided a distinction between the sheep experimentally infected with BSE and natural cases of sheep scrapie but did not provide a distinction between natural cases of bovine BSE and ovine scrapie. The sheep passaged CH1641 scrapie strain gave a glycoform ratio similar to ovine scrapie cases. The SSBP1 experimental scrapie strain gave a glycoform ratio which was different to that found for all the other samples (Fig. 2). The results suggested that this combination of molecular weight and glycoform ratio analyses, and differentiation with two specific antibodies could be used to provide a possible screening test to detect whether BSE has been naturally transmitted to sheep in the UK, if confirmed as accurate by bioassay and lesion profile analysis in mice inoculated with brain tissue from suspect field cases which is the present gold standard.

It is clear that Western immunoblotting is a versatile technique to study the prion protein both in terms of diagnosis, confirmation and some strain differ-

Figure 1 Western immunoblots using Protocol 5 – The VLA Hybrid technique used to discriminate between natural scrapie and experimental BSE in sheep samples.
Blot A represents samples processed with monoclonal antibody 6H4 (raised in mice to a bovine PrP amino acid sequence 144–152).
Blot B shows the results for the same samples using monoclonal P4 (raised in mice to an ovine PrP amino acid sequence 89–104).
Experimental BSE in sheep (lanes 1 and 2), natural cases of scrapie (lanes 3, 4, and 5) BSE in cattle (lane 6), M = Biotinylated molecular weight markers.

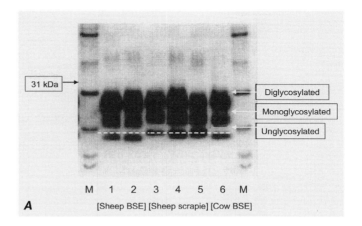

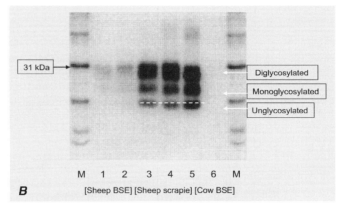

Blot A   Shows the characteristic three protein bands obtained for PrP^res associated with TSE diseases. It is evident that the electrophoretic mobility, particularly of the unglycosylated protein band is different, and dependent on whether the sample is of BSE or scrapie origin.
Blot B.   The same samples as Blot A, but processed using mAb P4. This antibody, as used in this particular test, is selective for scrapie PrP^res. BSE in sheep samples have very reduced levels of detection (lanes 1 and 2) the natural scrapie samples are well detected (lanes 3, 4 and 5) and the BSE in cattle sample is not detected at all (lane 6).

entiation. However, the techniques do have their inherent limitations with regard to speed and quantitative measurements and that is why a particular technique needs to be chosen depending on the particular study or survey requirements.

Applications of the different techniques have been highlighted where appropriate in the text. Troubleshooting information was given in text where appropriate.

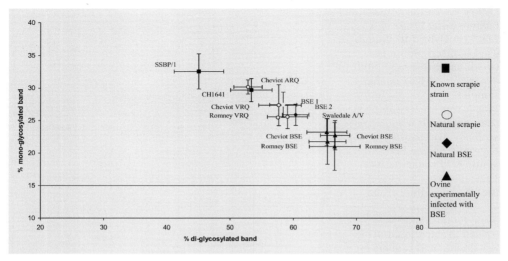

Figure 2 Scattergraph of the glycoform ratio of the proportion of abnormal protein in the di-glycosylated band and the mono-glycosylated band for natural UK bovine BSE cases (BSE1 and BSE2), natural UK scrapie cases of different genotypes* (Romney VRQ/VRQ, Cheviot VRQ/VRQ, Cheviot ARQ/ARQ, Swaledale ARQ/VRQ), two sheep-passaged scrapie strains (SSBP1 and CH1641) and ovines experimentally infected with BSE (Romney ARQ/ARQ BSE and Cheviot AHQ/AHQ BSE). Using the hybrid technique in combination with mAb 6H4 the SSBP1 and the experimentally infected ovine give unique glycoform ratios. There is considerable overlap of result between the natural cases of bovine BSE and the Romney, Cheviot VRQ/VRQ and Swaldale ovine scrapie. The CH1641 strain gives a closer ratio to that found for the Cheviot ARQ/ARQ scrapie sheep.

* Notes on genotype: The sheep PrP gene produces a protein of 256 amino acids, each of which is encoded by three DNA bases (one codon) in the gene. Susceptibility to scrapie has been shown to be linked to the PrP protein genotypes which are defined by variations in the amino acids encoded at codons 136, 154 and 171, and are termed polymophisms. At least five variant alleles have been found with respect to a risk of contracting scrapie which are depicted as ARQ, ARR, VRQ, AHQ and ARH. The codes represent polymorphism of amino acids at each codon i. e., $A_{136}R_{154}Q_{171}$ (ARQ) where A = alanine, R = argenine, Q = glutamine. The other two amino acids are; H = histidine and V = valine. Homozygous and heterozygous pairing of the two alleles inherited from a ram and a ewe therefore results in considerable variation of PrP genotype. This level of risk varies depending on breed type and the genotypes found within the flock.

Acknowledgements

The contributions and invaluable experience of my colleagues in the TSE Molecular Biology Department at VLA have helped provide this Chapter, they include; Melanie Chaplin, Jemma Clark, Linda Davis, and Jason Bramwell. My sincere thanks are also extended to Martin Groschup, at the Institute of Novel and Emerging Infectious Diseases, Insel Reims, Germany, who provided information for Protocols 1 and 2, Theodoros Sklaviadis from the Aristotle University of Thessaloniki, Greece, for Protocol 2, and Bruno Oesch and Marcus Moser, form Prionics AG, Zurich, Switzerland, for use of Protocol 4. Information for the increased sensitivity for Protocols 1 and 2 was provided from results obtained through an EU funded project FAIR PL 987021 which was very ably coordinated by Eddie Weavers.

References

1   Bolton DC, McKinley MP, Prusiner SB (1982) Identification of a protein that purifies with the scrapie prion. *Science* 218: 1309–1311

2   Chesebro B, Race R, Wherly K et al. (1985). Identification of scrapie prion protein specific mRNA in scrapie-infected and uninfected brain. *Nature* 315: 331–333

3   Oesch B, Westaway D, Walchli M et al. (1985). A cellular gene encodes scrapie PrP 27–30 protein. *Cell* 40: 735–746

4   Hunter GD, Millson GC, Meek G (1964) The intracellular location of the agent of mouse scrapie. *J Gen Microbiol* 34: 319

5   Millson GC, Hunter GD, Kimberlin RH (1971) An experimental examination of the scrapie agent in cell membrane mixtures. *J Comp Pathol* 81: 255–265

6   Prusiner SB, Garfin DE, Cochran SP et al. (1980) Experimental scrapie in the mouse: electrophoretic and sedimentation properties of the partially purified agent. *J Neurochem* 35: 574–582

7   Prusiner SB, Groth DF, Cochran SP et al. (1980a) Gel electrophoresis and glass permeation chromatography of the hamster scrapie agent after enzymic digestion and detergent extraction. *Biochemistry* 19: 4892–4898

8   Merz PA, Somerville RA, Wisniewski HM, Iqbal K (1981) Abnormal fibrils from scrapie-infected brain. *Acta Neuropath* 54: 63–74

9   Prusiner SB, Bolton DC, Groth DF et al. (1982) Further purification and characterisation of scrapie prions. *Biochemistry* 21: 6942–6950

10 Prusiner SB, Bolton DC, Kent SB, Hood LE (1984) Purification and structural studies of a major scrapie prion protein. *Cell* 38: 127–134

11 Hilmert H, Diringer H (1984) A rapid and efficient method to enrich SAF-protein from scrapie brains of hamsters. *Biosci Reps* 4: 165–170

12 Laemmli UK (1970) Cleavage of structural proteins during the assembly of the head of bacteriophage T4. *Nature* 277: 680–685

13 Towbin H, Staehlin T, Gordon J (1979) Electrophoretic transfer of proteins from polyacrylamide gels to nitrocellulose sheets: procedure and some applications. *Proc Nat Acad Sci USA* 76: 4350–4354

14 Collinge J, Sidle KCL, Meads J et al. (1996) Molecular analysis of prion strain variation and the aetiology of "new" Variant CJD. *Nature* 383: 685–690

15 Hill AF, Desbruslais M, Joiner S et al. (1997) The same prion strain causes vCJD and BSE. *Nature* 389: 448–450

16 Hope J, Reekie LJD, Hunter N et al. (1988) Fibrils from brains of cows with new cattle disease contain scrapie-associated fibrils. *Nature* 336: 390–392

17 Beekes M, Baldauf E, Caβens S et al. (1995) Western blot mapping of disease specific amyloid in various animal species and humans with transmissible spongiform encephalopathies using a high-yield purification method. *J Gen Virol* 76: 2567–2576

18 Scott AC, Stack MJ (1994) Protocols II and III. The diagnosis of bovine spongiform encephalopathy (BSE) and scrapie by the detection of BSE fibrils or scrapie-associated fibrils (SAF) by transmission electron microscopy and by Western Im-

munoblotting. In: European Commision for Agriculture: *Transmissible spongiform encephalopathies. Protocols for the laboratory diagnosis and confirmation of bovine spongiform encephalopathy and scrapie*. A report from the Scientific Veterinary Committee. European Commission, Directorate General for Agriculture, Unit for Veterinary Legislation and Zootechnics, Brussels, Prot. II: 1–12; Prot. III: 1–18

19 Stack MJ, Keyes P, Scott AC (1996) The Diagnosis of bovine spongiform encephalopathy and scrapie by the detection of fibrils and the abnormal protein isoform. In: HF Baker, RM Ridley (eds): *Methods in Molecular Medicine – Prion Diseases*. Humana Press, Totowa, N.J., 85–104

20 Benestad SL, Sarradin P, Thu B et al (2003) Cases of scrapie with unusual features in Norway and designation of a new type, Nor98. *Vet Rec* 153: 202–208

21 Polymenidou M, Verrghese-Nikolakaki S, Groschup M et al. (2002) A short purification process for quantitative isolation of PrP^{Sc} from naturally occurring and experimental transmissible spongiform encephalopathies. *BMC Infect Dis* 2: 23

22 Hadlow W, Kennedy RC, Race RE (1982) Natural infection of Suffolk sheep with scrapie virus. *J Infect Dis* 146: 657–664

23 Wilesmith J, Wells GAH, Cranwell MP, Ryan JBM (1988) Bovine spongiform encephalopathy: Epidemiological studies. *Vet Rec* 123: 638–644

24 Alpers M (1987) Epidemiology and clinical aspects of kuru. In: SB Prusiner, MP McKinley (eds): *Prions – novel infectious pathogens causing scrapie and Creutzfeldt-Jakob disease*. Academic Press, Orlando, 451–465

25 Hartsough GR, Burger D (1965) Encephalopathy of mink. 1. Epizootiologic and clinical observations. *J Infect Dis* 115: 387–392

26 Ikegami Y, Ito M, Isomura H et al. (1991) Pre-clinical and clinical diagnosis of scrapie by detection of PrP protein in tissues of sheep. *Vet Rec* 128: 271–275

27 Race RE, Ernst D (1992) Detection of proteinase K-resistant prion protein and infectivity in mouse spleen by 2 weeks after scrapie agent inoculation. *J Gen Virol* 73: 3319–3323

28 Rubenstein R, Merz PA, Kascsak RJ et al. (1991) Scrapie-infected spleens: analysis of infectivity, scrapie-associated fibrils, and protease-resistant proteins. *J Infect Dis* 164: 29–35

29 Van Keulen LJM, Schreuder BEC, Meloen RH et al. (1996) Immunohistochemical detection of prion protein in lymphoid tissues of sheep with natural scrapie. *J Clin Microbiol* 34: 1228–1231

30 Schreuder BEC, Van Keulen LJM, Vromans MEW et al. (1996) Preclinical test for prion diseases. *Nature* 381: 563

31 Hill AF, Zeidler M, Collinge J (1997) Diagnosis of new variant Creutzfeldt-Jakob disease by tonsil biopsy. *The Lancet* 349: 99–100

32 Kimberlin RH (1986) Scrapie: how much do we really understand? *Neuropath App Neurobiol* 12: 131–147

33 Kimberlin RH, Walker CA (1986) Pathogenesis of scrapie (strain 263K) in hamsters infected intracerebrally, intraperitoneally or intraocularly. *J Gen Virol* 67: 255–263

34 Kimberlin RH, Walker CA (1988) Incubation periods in six models of intraperitoneally injected scrapie depend mainly on the dynamics of agent replication within the nervous system and not the lymphoreticular system. *J Gen Virol* 69: 2953–2960

35 Kimberlin RH, Walker CA (1990) Intraperitoneal infection with scrapie is established within minutes of injection and is non-specifically enhanced by a variety of different drugs. *Arch Virol* 112: 103–114

36 Doi S, Ito M, Shinagawa M et al. (1988) Western blot detection of scrapie-associated fibril protein in tissues outside the central nervous system from preclinical scrapie-infected mice. *J Gen Virol* 69: 955–960

37 Mohri S, Farquhar CF, Somerville RA et al. (1992) Immunodection of a disease specific PrP fraction in scrapie-affected

sheep and BSE-affected cattle. *Vet Rec* 131: 537–539

38 Muramatsu Y, Onodera A, Horiuchi M (1994) Detection of PrP[Sc] in sheep at the preclinical stage of scrapie and its significance for the diagnosis of insidious infection. *Arch Virol* 13: 427–432

39 Race R, Ernst D, Jenny A et al. (1992) Diagnostic implications of detection of proteinase K-resistant protein in spleen, lymph nodes and brain of sheep. *Am J Vet Res* 53: 883–889

40 Safar J, Wille H, Itri V et al. (1998) Eight prion strains have PrP(Sc) molecules with different conformations. *Nat Med* 4: 1157–1165

41 Wadsworth JD, Joiner S, Hill AF et al. (2001) Tissue distribution of protease resistant prion protein in variant Creutzfeldt-Jakob disease using a highly sensitive immunoblotting assay. *Lancet* 358: 171–180

42 Kuczius T, Haist I, Groschup MH (1999) Molecular analysis of bovine spongiform encephalopathy and scrapie strain variation. *J Infect Dis* 178: 693–699

43 Kuczius T, Groschup MH (1999) Differences in proteinase K resistance and neuronal deposition of abnormal prion proteins characterise bovine spongiform encephalopathy (BSE) and scrapie strains. *Mol Med* 5: 406–418

44 Parchi P, Capellari S, Chen SG et al. (1997) Typing prion isoforms. *Nature* 386: 232–234

45 Somerville RA, Chong A, Mulqueen OU et al. (1997) Biochemical typing of scrapie strains. *Nature* 386: 564

46 Hill AF, Sidle KCL, Joiner S et al. (1998) Molecular screening of sheep for BSE. *Neurosci Lett* 255: 159–162

47 Hope J, Wood SCER, Birkett CR et al. (1999) Molecular analysis of ovine prion protein identifies similarities between BSE and an experimental isolate of natural scrapie, CH1641. *J Gen Virol* 80: 1–4

48 Hope J, Wood SCER, Birkett CR et al. (2000) Molecular analysis of ovine prion protein identifies similarities between BSE and an experimental isolate of natural scrapie, CH1641. *J Gen Virol* 81 (Pt 4): 1155–1164

49 Baron TGM, Madec J-Y, Calavas D (1999) Similar signature of the prion protein in natural sheep scrapie and bovine encephalopathy-linked diseases. *J Clin Microbiol* 37: 3701–3704

50 Baron TGM, Madec J-Y, Calavas D et al. (2000) Comparison of French natural scrapie isolates with BSE and experimental scrapie infected sheep. *Neurosci Lett* 284: 175–178

51 Sweeney T, Kuczius T, McElroy M et al. (2000) Molecular analysis of Irish sheep scrapie cases. *J Gen Virol* 81: 1621–1627

52 Stack MJ, Chaplin MJ, Clark J (2002). Differentiation of prion protein glycoforms from naturally occurring sheep scrapie, sheep passaged scrapie strains (CH1641 and SSBP1), bovine spongiform encephalopathy (BSE) cases and Romney and Cheviot breed sheep experimentally inoculated with BSE using two monoclonal antibodies. *Acta Neuropath* 104: 279–286

53 Nonno R, Esposito E, Vaccari G et al. (2003) Molecular analysis of cases of Italian sheep scrapie and comparison with cases of bovine spongiform encephaloapthy (BSE) and experimental BSE in sheep. *J Clin Microbiol* 41: 4127–4133

54 Schaller O, Fatzer R, Stack MJ et al. (1999) Validation of a Western immunoblotting procedure for bovine PrP[Sc] detection and its use as a rapid surveillance method for the diagnosis of bovine spongiform encephalopathy (BSE). *Acta Neuropathol* 98: 437–443

55 Moynagh J, Schimmel H (1999) Tests for BSE evaluated. *Nature* 400: 105

56 Foster JD, Dickinson AG (1988) The unusual properties of CH1641, a sheep-passaged isolate of scrapie. *Vet Rec* 123: 5–8

57 Dickinson AG, Fraser H (1979) An assessment of the genetics of scrapie in sheep and mice. In: SB Prusiner, WJ Hadlow (eds): *Slow Transmissible Diseases of the Nervous System, Vol 1*. Academic Press, New York, 367–386

58 Dickinson AG, Outram GW, Taylor DM, Foster JD (1986) Further evidence that scrapie agent has an independent genome. In: LA Court, D Dormont, P Brown, DT Kingsbury (eds): *Unconventional*

Viruses and Central Nervous System Diseases. Moisdon la Riviere, Abbaye de Mellaray, 446–460

59 Wilson DR, Anderson RD, Smith W (1950) Studies in scrapie. *J Comp Pathol* 60: 267–275

60 Dickinson AG, Outram GW (1988) Genetic aspects of unconventional virus infections: the basis of the vivino hypothesis. In: G Bock, J Marsh (eds): *Novel Infectious Agents and the Central Nervous System. Ciba Foundation Sympo-*

sium No 135. John Wiley & Sons, Chichester, UK, 63–83

61 Korth C, Stierli B, Streit P et al. (1997) Prion (PrP^{Sc}) – specific epitope defined by a monoclonal antibody. *Nature* 390: 74–77

62 Harmeyer S, Pfaff E, Groschup MH (1998) Synthetic peptide vaccines yield monoclonal antibodies to cellular and pathological prion proteins of ruminants. *J Gen Virol* 79: 937–945

Characterization of Anti-PrP Antibodies and Measurement of PrP using ELISA Techniques

Christophe Créminon and Jacques Grassi

Contents

Methods and Tools in Biosciences and Medicine
Techniques in Prion Research, ed. by S. Lehmann and J. Grassi
© 2004 Birkhäuser Verlag Basel/Switzerland

1   Introduction

Given the pivotal role played by PrP in the pathogenesis of transmissible spongiform encephalopathies (TSEs), the determination and characterization of the various forms of this protein (PrP^C and PrP^{Sc}) are key elements in prion research. The detection of PrP^{Sc} is also essential for the diagnosis of animal and human TSEs since this is the only unambiguous marker known for prion diseases. In this context, anti-PrP antibodies have proved to be precious tools for prion researchers. As an illustration, they are a key component in techniques described in four different Chapters in this book (see Chapters 7–9 of this volume). For more than 20 years, polyclonal and monoclonal antibodies have been produced in rabbits and mice by immunizing with PrP peptide, recombinant PrP or scrapie-associated fibrils (SAFs) in native or denatured form. Due the high conservation of the PrP sequence within mammalian species, PrP is usually poorly immunogenic in rodent species and the antibodies produced in normal animals are often characterized by a narrow species specificity. The most famous example is the monoclonal antibody (mAb) 3F4 raised in mice against hamster PrP that binds only human and hamster PrP due to the presence of a methionine residue in position 112, while it fails to bind mouse, sheep and bovine PrP which bear a valine residue at the same position [1–2]. The availability of PrP$^{0/0}$ mice offered a unique opportunity to raise antibodies with wider species specificity and directed against more uniformly distributed epitopes.

In recent years, several groups have described polyclonal or monoclonal antibodies specifically immunoprecipitating aggregated PrP^{Sc} as found in TSE-infected brains, in the form of SAFs. These antibodies were raised against synthetic peptides [3–4] recombinant PrP [5] or DNA [6] and are supposed to recognize either an epitope specifically exposed in PrP^{Sc} or DNA bound to PrP^{Sc} aggregates. To our way of thinking, the actual nature of the interactions involved in these immunoprecipitation processes is not clear and is still open to debate. It seems that the specificity of the immunoprecipitation is closely related both to the aggregation state of PrP^{Sc} and to the use of polydisperse solid phase as magnetic microbeads rather than an actual specific recognition by the corresponding antibodies. None of these works have led, so far, to the development of a diagnosis test with improved performances and the use of the corresponding antibodies is restricted to the group that have generated the reagent. From a practical point of view, it cannot be considered that antibodies specific for PrP^{Sc} are now available. As a consequence, the selective detection of PrP^{Sc} is still currently achieved under conditions taking advantage of the differences in biochemical properties between PrP^C and PrP^{Sc} (resistance to proteinase-K (PK) degradation, aggregation, cryptic epitopes) and makes use of antibodies directed against PrP^C or denatured PrP.

An important distinction between anti-PrP antibodies concerns the location and nature of the recognized epitope and the adequacy of binding properties of the antibody with the type of technique used. For instance, Western blot and immunohistochemistry, which are very popular methods in the prion research field, both involve denaturation of PrP. As a consequence, antibodies that give the best results with these techniques are essentially those recognizing linear continuous epitopes. The same holds for the detection of PrP^{Sc} which most of the time is associated with a denaturation step. In addition, since the specific detection of PrP^{Sc} is currently achieved after a treatment with PK, intended to eliminate PrP^{C} immunoreactivity in the sample, and removing the N-terminal part of PrP^{Sc}, it is generally recommended to use for this purpose antibodies directed against the C-terminal part (downstream of residue 92). In contrast, other techniques such as immunocytochemistry, immunoprecipitation, ELISA and flow cytometry may require recognition of native PrP. In this case, good results are more often achieved by using antibodies binding conformational discontinuous epitopes.

In this paper, we shall describe ELISA techniques suitable for characterizing anti-PrP polyclonal or monoclonal antibodies as well as sandwich immunoassays allowing sensitive measurement of the various forms of PrP.

2 Materials

Chemicals
Biotin N-hydroxysuccinimide ester (SIGMA # H 1759); biotin-maleimide (SIGMA # B 1267); acetylthiocholine iodide (SIGMA # A 5751); 5,5'-dithiobis(2-nitrobenzoic acid) (SIGMA # D 8130); bovine albumin (SIGMA # A7906); Igepal CA-630 (SIGMA # I 3021); sodium deoxycholate monohydrate (SIGMA # D 5670), Thimerosal (SIGMA # 8784).

Antibodies
Anti-PrP antibodies are now available from several commercial sources. Some of them are listed below (non-exhaustive list):
- SPI-Bio, 2 rue du Buisson aux Fraises, Z.I de la Bonde, 91741 Massy cedex, France. Tel: +33-1-63-9-53-14-00, Fax: 33-169-53-15-00, Contact: contact@spibio.com
- Prionics, Wagistrassse 27a, Ch-8952 Shlieren, Switzerland. Tel: +41-44-200-20-00, Fax: +41-44-200-20-10, Contact: info@prionics.ch
- Roboscreen GmbH, Delitzscher Strasse 135 D-04129 Leipzig, Germany. Tel: +49-341-9725970, Fax: +49-341-9725979, Contact: aosman@robocreen.com
- VMRD, Inc.P.O. Box 502Pullman, WA 99163, USA, Tel: +1-509-334-5815, Fax: +1-509-332-5356, Contact: vmrd@vmrd.com

- R-Biopharm AG, Landwehrstr. 54, 64293 Darmstadt, Germany, Tel: +49-6151-8102-0, Fax: +49-6151-8102-40, Contact: info@r-biopharm.de
- DakoCytomation Denmark, Produktionsvej 42, DK-2600 Glostrup, Denmark, Tel: +45-44-85-95-00, Fax +45-44-85-95-95, Contact: contact@dakocyto-mation.com
- alicon AG, Wagistrasse 23, 8952 Zürich-Schlieren, Switzerland, Tel.: +41 43 495 05 67, Fax: +41 43 495 05 69, Contact: info@alicon.ch

Secondary antibodies: In the examples presented in this paper, we make use of anti-mouse immunoglobulin antibodies produced by the Jackson company: Affipure goat anti-mouse IgG+IgM (H+L) antibodies: Jackson ImmunoResearch Laboratories Inc (# 115-005-044), West Grove, PA 19390, USA. However, similar reagents can be obtained from other sources including SIGMA-Aldrich (St Louis, USA) and PIERCE (Rockford, USA).

Recombinant PrP can be obtained from various sources including Prionics, alicon AG and Roboscreen (see above).

Equipment
- 96-well microtitre plates (Immunoplate Maxisorp with certificate) Nunc, Denmark.
- Plate washer ELX 405 (BioTek Instruments, USA).
- Automatic plate distributor Multidrop and plate reader Multiskan EX from Labsystems, Finland.
- A Ribolyser type tissue homogeniser, Thermo-Hybaid (Milford MA, USA).

Solutions, reagents and buffers
- Solution 1 – EIA buffer: 100 mM potassium phosphate buffer pH 7.4 containing 0.1% BSA, 150 mM sodium chloride and 0.01% sodium azide.
- Solution 2 – Washing buffer: 10 mM potassium phosphate buffer pH 7.4 containing 0.05% Tween 20.
- Solution 3 – Ellman medium: 0.75 mM acetylthiocholine and 0.5 mM 5,5′-dithiobis(2-nitrobenzoic acid) in 100 mM phosphate buffer pH 7.4.
- Solution 4 – Lysis buffer: 10 mM Tris/HCl buffer pH 7.4 containing 100 mM NaCl, 10 mM EDTA, 0.5% Igepal CA-630 (NP-40) and 1% sodium deoxycholate.

3   Methods

Notes: All the procedures described in this Chapter have been especially optimized in our laboratory using acetylcholinesterase (AChE)-labelled enzyme conjugates, because this enzyme is particularly well suited to these applications. However, the same principles can be applied, with appropriate modifications, using other enzyme conjugates (peroxidase, alkaline phosphatase and β-galactosidase) available from numerous commercial sources. Some recommendations concerning the adaptation required when peroxidase conjugates are used are given in the troubleshooting section (see section 4 of this Chapter). AChE concentrations are given in terms of Ellman's units (EU); one EU corresponds to 8 ng of enzyme (25 fmol of the tetramer).

3.1   Screening for anti-PrP antibodies

General considerations

Screening methods are critical in the production of anti-PrP antibodies. The final evaluation of the properties of an antibody in a given technique (Western blot or immunohistochemistry (IHC), for instance) can be established only by testing the antibody under the real conditions of the technique. This is the purpose of what we call "secondary screening". However, in many situations, a primary screening, using a simple and more rapid technique, is very useful. This is particularly true when producing monoclonal antibodies since in most cases hundreds or thousands of hybridoma culture supernatants have to be analyzed. From a general point of view, an ideal primary screening method should check the capacity of an antibody to bind the immunizing preparation (here PrP peptides, recombinant PrP or SAFs). This can be achieved very efficiently by using ELISA techniques performed in polystyrene 96-well microtitre plates. Basically two contrasting strategies can be used. In the first one (direct ELISA), the antigen is coated on the plastic and antibodies bound to solid phase-immobilized antigen are detected using an anti-immunoglobulin (anti-Ig) antibody labelled with an enzyme. The alternative strategy makes use of the same components but reverses the orientation of the reactants (indirect ELISA). It involves the coating of the anti-Ig in the wells and the visualization of solid-phase bound antibodies by reaction with an enzyme-labelled antigen. The first strategy is certainly the more widely used because it is very simple to perform and requires only the coating of the antigen and the use of commercially available secondary antibodies. In contrast, the labelling of the antigen, required for indirect ELISA, is usually considered more difficult to achieve. However, we recommend the second strategy because it has long been documented that coating of protein on plastic is a highly denaturing process that induces profound modifications of the protein conformation which may lead to a loss of immunoreactivity towards antibodies recognizing conformational/discontinuous epitopes. However, since in some situations it is interesting to select antibodies recognizing the denatured form of PrP (antibodies for Western blot applications or antibodies recognizing denatured PrP^{Sc}), we shall describe the two approaches in detail.

Protocol 1   Coating of PrP or of an anti-Ig antibody on microtitre plates

Coating of a protein on a polystyrene surface involves a passive adsorption of the protein and is easily achieved by direct contact of a solution of pure protein with the plastic. The same procedure is used to immobilize anti-Ig or PrP. It is essential to use specific polystyrene plates optimized for this purpose (immunology quality) and not plates intended for the culture of cells or tissues. Anti-Ig antibodies (polyclonal or monoclonal) are commercially available from numerous companies (see Materials). The procedure described here applies to the coating of recombinant PrP, which is now available from various companies (see Materials), but the same technique can be used to immobilize purified PrP^C. Once the coating of PrP or anti-Ig is accomplished, the solid phase is saturated by reaction with a buffer containing a high concentration of bovine serum albumin (BSA) in order to reduce non-specific binding.

1. Prepare a 5 µg/ml solution of PrP or second antibody by diluting the corresponding stock solution of the protein in 50 mM phosphate buffer pH 7.4.
2. Add 100 µl of the prepared solution to each well of the microtitre plate, using either a manual multichannel micropipette or a Combitips dispenser or an automatic device, for 18 h reaction at 20 °C.
3. Wash the plates with washing buffer (Solution 2) (three cycles of aspiration and dispensing) before filling the wells with 300 µl of EIA buffer (Solution 1) to ensure total saturation of the polystyrene surface.
4. Seal the coated microtitre plates with an adhesive cover sheet and store at +4 °C.
   Notes: These plates can be used 18 h after performing the saturation step and are stable for at least six months when stored at 4 °C.

Protocol 2   Labelling of PrP or PrP peptides with biotin

The labelling of protein or peptides with enzymes is currently achieved by covalent coupling. However, this requires quite sophisticated chemistry which is not easily accessible to all research laboratories. This difficulty can be overcome by taking advantage of the remarkable properties of the high-affinity avidin–biotin interaction which provides an easy way to bind an enzyme to any kind of protein or peptides. In this case the only requirement is to label the protein with biotin, which is easily accomplished using commercially available reagents. The binding with an enzyme will be achieved in a second step by reaction with enzyme-labelled avidin. The coupling of a biotin molecule to a protein or a peptide is obtained by reaction of the N-hydroxysuccinimide ester of biotin (NHS-biotin) with primary amino groups of the protein or peptide, in mild alkaline conditions. Alternatively, this can also be achieved using a maleimide derivative of biotin further reacting with the thiol group(s) of the protein or peptide.

Notes: NHS-biotin is currently used in large excess because the yield of the reaction is very low (most of NHS-biotin is hydrolysed by water). However, due to the presence of numerous primary amino functions into the PrP sequence (10 or 11 depending on the species), we have to avoid excess incorporation that could lead to further alteration of the recognition by antibodies. The NHS-biotin/recPrP ratio used during this procedure thus results from a compromise between these two contradictory requirements.

1. Dissolve 100 µg of recPrP, corresponding to about 4 nmol, in 500 µl of 100 mM borate buffer pH 8.5.
2. Add 10 µl of a 3.41 mg/ml solution of NHS-biotin in N,N-Dimethyl-formamide (DMF), corresponding to 100 nmol.
3. Mix rapidly and let the reaction proceed for 1 h at 20 °C.
4. Add 100 µl of 1 M Tris-HCl pH 8 to neutralize the remaining active ester.

Notes: Excess biotin is separated by means of molecular sieve chromatography.

5. The biotin-PrP is purified using a Sephadex G-25 column (1.5 × 250 mm) equilibrated and eluted with 100 mM phosphate buffer pH 7.4.
6. Collect 1 ml fractions and measure their absorbance at 280 nm.

Notes: The biotin-PrP conjugate is eluted as a single peak in the void volume of the column.

7. Pool the fractions corresponding to the biotin-PrP conjugate and dilute with an equal volume of EIA buffer (Solution 1) to avoid further possible loss due to adsorption.

Notes: When PrP peptides are concerned, the reaction generally proceeds via the same strategy involving NHS-biotin and the α-amino function of the peptide.

1. Dissolve 1 µmol of the peptide in 500 µl of 100 mM borate buffer pH 8.5.
2. Add 100 µl of 34.41 mg/ml solution of NHS-biotin dissolved in DMF (10 µmol) for 1 h reaction at 20 °C.

Notes: When the peptide contains several lysine residues (sequence 100–110 of the mouse PrP for instance), the use of the excess of NHS-biotin described above may result in simultaneous incorporation of several biotins, possibly leading to a biotin-peptide conjugate unreactive towards antibodies. The reaction must then be performed using a lower NHS-biotin/ peptide ratio, i. e., 3 molar excess.

Alternatively, it may be more relevant to incorporate during peptide synthesis an extra cysteinyl residue (either at the amino- or the carboxy-terminus of the peptide) which can specifically and very efficiently react with a maleimido-biotin derivative.

1. Dissolve 250 nmol of the peptide containing the extra cysteinyl residue in 500 µl of 100 mM phosphate buffer pH 6.5.
2. Add 2.5 µmol (1.128 mg) of maleimido-biotin dissolved in 50 µl of DMF for 1 h of reaction at 30 °C.

Notes: The further purification of the biotin-peptide conjugate from the excess biotin reagent can be achieved by molecular sieve chromatography. The reaction mixture is chromatographed using a Biogel PD2 or Sephadex G15 column (15 × 250 mm) equilibrated and eluted as specified above. However, a correct separation is sometimes difficult to achieve by this means since the molecular weights of the conjugate and contaminants are not significantly different. To obtain better results, another chromatography method based either on hydrophobicity, such as solid-phase extraction, or on the charge of the products (ion-exchange chromatography) should then be envisaged.

In some situations, it is not necessary to eliminate the excess of unreacted biotin by using chromatographic techniques. This will be done during the course of the immunoassay during washing steps. This is the case for the assay described in Protocol 4.

Protocol 3   Detection of polyclonal or monoclonal antibodies using immobilized PrP

In this procedure the detection is done in two steps. In the first step, the solution containing the antibody (serum dilution or hybridoma culture supernatant) is reacted with solid-phase immobilized PrP. In the second step, the presence of the antibody on the solid phase is revealed by reaction with an enzyme-labelled anti-Ig antibody.

1. Wash the plates previously coated with 100 µl of the PrP solution (three cycles).

Notes: To analyze and compare the polyclonal response from different immunized animals, it is useful to prepare serial dilutions (1/100, $1/10^3$, $1/10^4$, $1/10^5$) of the sera in EIA buffer. In contrast, when screening hybridoma supernatants, the situation is rather different since the goal is to identify those supernatants containing specific anti-PrP antibodies among a large number of negative supernatants (usually 1,000 to 3,000). In this case, undiluted culture supernatant will be tested. Since both types of microplate, i. e., cell culture and coated plates, share the same geometry, this operation can be easily and rapidly performed using a multi-channel micropipette.

2. Dispense either 100 µl of the antisera dilutions in two consecutive wells in order to perform a duplicate analysis or 100 µl of the culture supernatants into the wells containing solid phase-immobilized PrP.

Notes: To improve the analysis of the results, control wells for measurement of non-specific binding must be included in the experiment.

3. Non-specific binding: dispense 100 µl of either EIA buffer or culture medium into at least two wells.

4. Seal the plates using an adhesive plastic cover sheet to avoid evaporation.

5. Let the immunoreaction proceed for at least 4 h at 20 °C or preferably 18 h at 4 °C.

6. Wash the plates extensively using washing buffer (Solution 2, at least 2 × 3 cycles, including a 3 min soaking step between the 2 cycles of 3).

7. Add 100 µl/well of enzyme conjugate, i.e., AChE-labelled sheep polyclonal anti-mouse Ig or mouse monoclonal anti-rabbit Ig antibodies depending on the species analyzed, at a 2 EU/ml concentration.
8. Incubate for 4 h at 20 °C.
9. Wash the plates extensively as in 3.6.
10. Add 200 µl/well of Ellman medium.
11. Let the enzymatic reaction proceed at 20 °C for 30 min to 1 h, e.g., when a yellow colour appears in the wells, but longer reaction times can be used.
12. Measure absorbance at 414 nm.

Protocol 4   Detection of polyclonal or monoclonal antibodies using PrP or PrP peptides labelled with biotin

In this procedure, the solution containing the antibody (serum dilution or hybridoma culture supernatant) is first reacted with a solid phase containing an anti-Ig antibody (see Protocol 1) and the presence of anti-PrP antibodies on the solid phase is revealed by reaction with biotin-labelled PrP. In a final step, enzyme-labelled avidin is added in order to detect the biotinylated PrP bound to the solid phase. In this format, the binding of antibodies to solid phase-immobilized anti-Ig and the reaction of anti-PrP with biotin-PrP can be achieved in a single step.

1. Wash the plates coated with a secondary antibody (anti-rabbit or mouse Ig antibody) (three cycles).
2. Dispense 50 µl of a solution of rabbit antiserum or hybridoma culture supernatant into the wells as described in 3.2.
3. Add 50 µl of biotin-labelled PrP or peptide diluted in EIA buffer at a concentration ranging between 10 and 100 ng/ml.
4. Evaluate the non-specific binding of the tracer by including two wells corresponding to the incubation of 50 µl of EIA buffer or culture medium (as described in 3.3) with 50 µl of the biotin-labelled molecule.
5. Seal the plates with an adhesive cover sheet.
6. Incubate for 18 h at 4 °C (or 4 h at 20 °C).
7. Wash the plates (2 × 3 cycles including a 3 min soaking step).
8. Add 100 µl of enzyme conjugate, i.e., streptavidin-AChE at 2 EU/ml.
9. Let the immunoreaction proceed for at least 2 h at 20 °C.
10. Wash the plates (2 × 3 cycles including a 3 min soaking step).
11. Dispense 200 µl Ellman medium/well.
12. Measure absorbance at 414 nm after 30 min to 1 h of enzymatic reaction, depending on the intensity of the yellow colour.

3.2 Detection and measurement of PrP using ELISA techniques

General considerations

Basically the detection of PrP using ELISA techniques can be achieved in two ways. In a first approach, PrP is coated on the plastic before being reacted with an enzyme-labelled antibody (direct ELISA). In a second approach, PrP is specifically captured on a coated antibody and further detected by reaction with a second antibody labelled with an enzyme (sandwich ELISA). Although direct ELISA successfully detects PrP^{Sc} (see application), the sandwich ELISA is strongly recommended because it ensures more specific and consequently more reliable capture. In addition, it has been widely documented that the sandwich ELISA (also called two-site immunometric assay) is the format offering the most sensitive detection for antigens [7].

Protocol 5   Labelling of anti-PrP antibodies with biotin

As already mentioned in Protocol 2, the most efficient way to prepare an enzyme-labelled antibody is covalent coupling, but for the sake of simplicity it is possible to resort to biotin labelling as described above. The labelling of an antibody with biotin is done using a procedure very similar to the one described previously for PrP (see Protocol 2).

1. In a 5 ml polypropylene tube, dissolve 450 µg of anti-PrP monoclonal antibody, grossly corresponding to 3 nmol, in 500 µl of 100 mM borate buffer pH 8.5.
2. Add 20 µl of a 1.28 mg/ml NHS-biotin solution in DMF (corresponding to an NHS-biotin/protein ratio of 25) and mix thoroughly.
3. Allow to react for 1 h at 20 °C.
4. Add 200 µl of 200 mM Tris buffer pH 8 for 15 min.
5. Complete with 3.78 ml of EIA buffer.

Notes: Steps 4 and 5 are designed to stop the reaction by providing excess amino functions and to prevent adsorption of the biotinylated antibodies by addition of a protein buffer.

6. Aliquot the resulting stock solution (100 µg/ml) of the conjugate and store frozen at –20 °C before use.

Notes: biotin-labelled antibody can be purified using a Sephadex G-25 column (1.5 × 250 mm) equilibrated and eluted with 100 mM phosphate buffer pH 7.4 as described in Protocol 2.

Protocol 6   Measuring PrP^C in tissues using a sandwich immunoassay

Since PrP^C is usually linked to cell membranes via a glycosyl-phosphatidyl-inositol (GPI) anchor, its measurement first requires an extraction process using detergents to recover "soluble" PrP. The only exception is PrP^C present in serum, plasma, urine or milk which exists as a soluble protein and can thus be measured directly. PrP^C can be extracted from tissues by a large variety of detergents or detergent mixtures. Usually, the extraction by itself is preceded by homogenization of the tissue in the absence of detergent to avoid the formation of foam. For instance, a 10% or 20% (w/v) tissue homogenate can be prepared in a 5% glucose solution. Extraction of PrP will be achieved in a second step by adding a detergent solution followed by strong mixing (vortexing) of the mixture. For ELISA applications, it is not recommended to use strong ionic detergents like N-lauryl sarcosyl or sodium dodecyl sulphate which are hardly compatible with antibody binding. Detergents like Triton X100 (1% or 2%) or a mixture of Triton X100/sodium deoxycholate or Nonidet NP40/sodium deoxycholate should be preferred. A typical extraction scheme suitable for a subsequent ELISA measurement has been described in Moudjou et al. [8].

1. Homogenize the tissues (20% w/v) in 5% glucose using a Ribolyzer.
2. Further extract the PrP^C by adding 3 volumes of lysis buffer (Solution 3) and vortex.
3. Centrifuge for 5 min at 1,000 rpm.
4. Discard the pellet, aliquot the supernatants and store at −20 °C until analysis of the PrP^C content.

Notes: Extracts from the different tissues can be analyzed as serial dilutions in EIA buffer (preferably containing 0.1% Triton X100). Recombinant PrP purified and diluted in the same buffer can be used to generate a standard curve in the 100 ng/ml to 100 pg/ml range (depending on the capture and tracer antibody characteristics).

5. Wash the plates coated with a monoclonal anti-PrP antibody (three cycles).
6. Dispense 100 µl in duplicate for each point of the standard curve or extract dilutions.
7. Evaluate non-specific binding of the tracer (in at least 4 wells) by incubating 100 µl of the buffer under the same conditions.

Notes: If available, the best negative control is the corresponding extract from PrP$^{0/0}$ mouse tissues which can be used to check the influence of possible non-specific interference in the sandwich assay.

8. React for 2 or 3 h at 20 °C.
9. Wash the plates (three cycles).
10. Add 100 µl/well of biotin-labelled tracer antibody (10 to 100 ng/ml diluted in EIA buffer).
11. Seal the plates with an adhesive cover sheet.
12. React for 2 or 3 h at 20 °C.
13. Wash the plates (2 × 3 cycles including a 3 min soaking step).
14. Add 100 µl of enzyme conjugate, i. e., streptavidin-AChE at 3 EU/ml.

15. React for at least 2 h at 20 °C.
16. Wash the plates (2 × 3 cycles including a 3 min soaking step).
17. Dispense 200 µl/well of Ellman medium.
18. Allow the enzymatic reaction to proceed at 20 °C for 30 min to 1 h before reading, e. g., when a yellow colour appears in the wells, but longer reaction times can be used.
19. Measure absorbance at 414 nm.
20. Results can be used in terms of absorbance measurements and will then provide a relative determination of PrP^C. If a standard curve has been generated with known amounts or concentrations of recombinant PrP, results can be expressed in absolute amounts or concentration. However, these absolute results should be considered with caution if it is not demonstrated that recombinant PrP and PrP^C have the same immunoreactivity with regard to the two antibodies used in the sandwich immunoassay. The quality of the recombinant PrP preparation is also essential for a good standardization of the assay.

Notes: the duration and temperature given in this "typical" procedure for the immunological reaction steps (steps 8 and 12) are only indicative and have to be optimized for each individual assay depending on the characteristics of the antibodies used. For instance, in many situations, a more sensitive detection will be obtained using overnight reaction at +4 °C, but with a significant increase in the total duration of the assay. The same holds for the concentration of the tracer antibody which must be adjusted so that an optimal compromise is found between non-specific binding and the intensity of the PrP-related signal.

Protocol 7   Measuring PrP^{Sc} in tissues using a sandwich immunoassay

In the absence of antibodies specifically recognizing PrP^{Sc}, and since PrP^C is always present in tissues accumulating PrP^{Sc}, its measurement requires both the destruction of PrP^C by a PK treatment and a further denaturation of PrP^{Sc} so that antibodies recognizing PrP^C or denatured PrP can be used. The PK treatment is usually applied to a tissue homogenate or extract and is performed in the presence of detergents as described in several chapters in this book (see Chapters 3, 7, and 8 of this volume). In many situations it is preferable to prepare scrapie-associated fibrils as described in Chapters 3 and 8. The choice of the denaturing process is also critical because the denaturing of PrP^{Sc} must be done under conditions compatible with subsequent detection of denatured PrP by a sandwich ELISA. As mentioned above, strong ionic detergents as N-lauryl sarcosyl or sodium dodecyl sulphate should be omitted. In our experience, the best compromise is to use highly concentrated chaotropic agents like urea (4–6 M), guanidine (4–8 M) or guanidine thiocyanate (4 M). Denaturation is achieved by heating at 100 °C for 10 min. After denaturation, the PrP solution has to be diluted in an appropriate buffer before analysis by sandwich ELISA.

1. Mix the PrP^{Sc}-containing solution (v/v) with 8 M urea, leading to a final 4 M urea concentration.
2. Heat at 100 °C for 10 min to denature PrP^{Sc}.
3. Dilute each tissue sample at least 4–8-fold in EIA buffer before performing the sandwich assay.

Notes: the dilution factor essentially depends on the properties of the capture antibody whose binding may be incompatible with high concentrations of urea.

4. Prepare serial dilutions of the resulting solution containing denatured PrP^{Sc} in EIA buffer.

These serial dilutions can be further assayed by following a protocol strictly similar to steps 5 to 20 of the Protocol 6 for PrP^{C} measurement (possibly using the same recombinant protein as standard).

4   Troubleshooting

Most technical details and troubleshooting have been mentioned in the text as 'Notes'. However, some additional recommendations can be given:

1. The performance of the microtitre plates containing an immobilized biological material (either antigen or antibody) is profoundly altered when the solid phase is brought to dryness. As a consequence, it is critical to avoid any drying of the plates, which should be washed for just a few minutes before the addition of the reactants.
2. As far as possible, all dilutions performed on biological material (tissue extract, antibody or antigen solutions) should be done in a protein-containing buffer (for instance, EIA buffer, Solution 1, given in this Chapter) in order to avoid any loss of protein due to absorption on vessel walls.
3. The properties of antigen or antibody solutions can be significantly altered by several cycles of freezing and thawing. As a consequence, the corresponding solutions should be aliquoted (for instance as the amount corresponding to one microtitre plate) and kept frozen at –20 °C or (preferably) at –80 °C.
4. If peroxidase-labelled reagents are used, it is absolutely necessary to remove any trace of sodium azide in all buffers and solutions because this compound irreversibly inhibits the enzyme activity. Sodium azide should be replaced by thimerosal (see Materials).
5. The quality of ELISA measurements is closely related to the quality of pipetting and washing steps of the assay. Precautions must be taken to avoid contamination of solutions with highly concentrated reagents (stocks solutions of tracer antibody, for instance). This can be done by changing pipette tips as often as possible.

5   Applications and conclusion

During the 25 last years ELISA techniques have proved to be a method of choice for quantitative measurement of almost every kind of biological molecule, and they are widely used in all areas of biological and medical research. There is no doubt that they are also very useful in the field of prion research. They provide a very efficient way to characterize the binding properties of polyclonal and monoclonal antibodies with regards to the various forms of PrP (PrP peptide, recombinant PrP, PrP^C and PrP^{Sc}) and are thus very useful as primary screening methods. When compared to Western blot analysis, which is certainly the most popular method in this particular field, ELISA offers the possibility of a more rapid and easier determination of the various forms of PrP with the possibility of a quantitative measurement. This is certainly of great interest for every kind of basic research, but the most obvious application concerns the diagnosis of TSEs in animals or humans. This is illustrated by the fact that out of five tests officially approved by the European Community for the post-mortem diagnosis of BSE in cattle [9], four are ELISA-related techniques (Enfer, Bio-Rad, Prionics LIA and UCSF-InPro) with only one Western blot method (Prionics check). As mentioned earlier in this document, all these tests are based on the immunological detection of PrP^{Sc}. Most make use of PK treatment, in order to eliminate PrP^C in brain samples, and all involve a denaturing step allowing the detection of denatured PrP^{Sc} by antibodies. One of these tests (Enfer) is a direct ELISA which involves the immobilization of PrP^{Sc} on plastic, while the three other tests (Bio-Rad, Prionics LIA, and the UCSF–InPro test) are based on the use of a sandwich immunoassay. In addition, two of these rapid tests include a purification/concentration step which precedes the immunological detection of PrP^{Sc} (Bio-Rad, UCSF–InPro). This certainly increases the analytical sensitivity of the corresponding diagnosis test as shown in comparative validation studies performed under the control of the European Commission [9]. However, the intrinsic qualities of the antibodies used for the detection are still critical and we can assume that the next generation of tests will include ELISA-based techniques.

Further reading

JD Pound (ed) (1998) *Immunochemical Protocols*. Humana Press Inc., New Jersey, USA

CP Price, DJ Newman (eds) (1997) *Principles and Practice of Immunoassay* (second edition). Macmillan Reference Ltd., New York, USA

References

1   Kascsak RJ, Rubenstein R, Merz PA et al (1987) Mouse polyclonal and monoclonal antibody to scrapie-associated fibril proteins. *J Virol* 61: 3688–3693

2   Bolton DC, Seligman SJ, Bablanian G et al (1991) Molecular location of a species-specific epitope on the hamster scrapie agent protein. *J Virol* 65: 3667–3675

3   Paramithiotis E, Pinard M, Lawton T et al (2003) A prion protein epitope selective for the pathologically misfolded conformation. *Nat Med* 9: 893–899

4   Serbec VC, Bresjanac M, Popovic M et al (2003) Monoclonal Antibody against a Peptide of Human Prion Protein Discriminates between Creutzfeldt-Jacob's Disease-affected and Normal Brain Tissue. *J Biol Chem* 279: 3694–3698

5   Korth C, Stierli B, Streit P et al (1997) Prion (PrP^{Sc})-specific epitope defined by a monoclonal antibody. *Nature* 390: 74–77

6   Zou WQ, Zheng J, Gray DM et al (2004). Antibody to DNA detects scrapie but not normal prion protein. *Proc Natl Acad Sci USA* 101: 1380–1385

7   Ekins RP, Jackson T (1984) Non isotopic immunoassays. An overview. In: CA Bizollon (ed) *Monoclonal antibodies and new trends in immunoassay*. Elsevier, Amsterdam, 149–163

8   Moudjou M, Frobert Y, Grassi J, La Bonnardiere C (2001) Cellular prion protein status in sheep: tissue-specific biochemical signatures. *J Gen Virol* 82: 2017–2024

9   Moynagh J, Schimmel H (1999) Tests for BSE evaluated. Bovine spongiform encephalopathy. *Nature* 400: 105

Further information is available at: http://europa.eu.int/comm/food/fs/bse/bse12_en.pdf

and http://europa.eu.int/comm/food/fs/bse/bse42_en.pdf

10 TSE Strain Typing in Mice

Moira E. Bruce, Aileen Boyle and Irene McConnell

Contents

1 Introduction

It has been known for many years that transmissible spongiform encephalopathy (TSE) agents exhibit strain variation [1], a phenomenon that has been studied most extensively in experimental mouse models. Numerous distinct TSE strains have been isolated in mice from a range of scrapie, bovine spongiform encephalopathy (BSE) and Creutzfeldt-Jakob disease (CJD) sources. The methods used for TSE strain discrimination are based on simple observations of

Methods and Tools in Biosciences and Medicine
Techniques in Prion Research, ed. by S. Lehmann and J. Grassi
© 2004 Birkhäuser Verlag Basel/Switzerland

disease characteristics, the most useful being the length of the incubation period of the disease and the type and extent of the pathological changes seen in the brains of infected animals [24]. Formal strain typing protocols in mice, based on incubation periods and neuropathology, have been used extensively as tools in basic research, for example in studies exploring the nature of TSE agents.

TSEs in non-transgenic mice have long asymptomatic incubation periods, lasting approximately between four months and the full lifespan of the mouse (over two years), depending on the model. Following this asymptomatic phase, progressive neurological signs are seen, leading to terminal disease, usually within a few weeks. Despite this protracted time-course, the incubation period is remarkably predictable. For example, a single TSE strain injected intracerebrally at high dose into a group of genetically uniform mice will generally give a mean incubation period with a standard error of less than 2% of this mean. The incubation period is also highly predictable for different groups of genetically uniform mice injected with equivalent doses of the same TSE strain. However, markedly different incubation periods are seen when different TSE strains are tested in a single inbred mouse strain [4].

The incubation period is also profoundly influenced by host genetic factors [2], associated mainly with variation in the gene encoding the prion protein (PrP). In mice, two alleles of this gene (*Prn-p*) have been recognised (designated a and b), encoding proteins that differ by two amino acids, at codons 108 and 189 [5]. For a single TSE strain, the incubation periods in mice of different PrP genotypes may differ by hundreds of days, an effect that is related to the rate of progression of the disease, rather than to differing susceptibilities to infection. The effects of PrP genotype on disease progression were identified many years ago, long before the protein itself was discovered [6]. The gene that was later found to encode PrP [7, 8] was originally called *Sinc* (acronym for scrapie incubation), with the two *Sinc* alleles s7 and p7 corresponding to the a and b alleles of the PrP gene. Genes other than the PrP gene may also influence incubation period, but usually to a lesser extent.

The incubation period of the disease therefore depends on a precise interaction between the TSE strain and host PrP, with each TSE strain producing a characteristic and highly reproducible pattern of incubation periods in the three possible PrP mouse genotypes (the two homozygotes and the heterozygote F_1 cross) [2, 4]. The molecular basis of this interaction is still not understood.

TSE strains also vary dramatically in the spectrum of pathological changes that they produce in the brains of infected mice [9]. In routine histological sections, TSE-specific vacuolation is seen to be targeted to different brain areas, depending mainly on the TSE strain, but also to some extent on PrP and other host genetic factors. This is the basis of a semi-quantitative method of strain discrimination in which the severity of vacuolation is scored from coded sections in nine grey matter and three white matter brain areas to construct a 'lesion profile' that is characteristic for each combination of TSE strain and inbred mouse strain [3,4]. More detailed studies have shown that vacuolation and associated PrP accumulation are targeted precisely to anatomically defined brain areas [10]. This suggests that a fundamental difference between TSE strains is their ability to recognize and replicate in different neuronal populations, but the molecular basis of this selectivity is not yet understood.

The combination of incubation period and lesion profile data generated in mice acts as a "signature" for the TSE strain. It is now established that these methods can be used to type the TSE strain present in a naturally infected host, to explore epidemiological links between TSEs occurring in individuals of the same or different species. For example, the same TSE strain has been detected in tissues from BSE cattle and from several other species with novel TSEs, including patients with variant CJD (vCJD) [11–13].

This Chapter documents the standard TSE strain typing protocol, as used for many years at the Neuropathogenesis Unit (NPU) of the Institute for Animal Health. Although details of the protocol are to some extent arbitrary, e. g., the use of particular mouse strains, standardisation of the protocol has allowed direct comparisons between a large number of different TSE isolates. Similar approaches are being used elsewhere, with variations to this protocol.

2   Materials/Facilities

Mouse strains
- C57BL/Dk: a substrain of C57BL maintained at the NPU as an inbred line for over 35 years, carrying the a allele of the PrP gene (*Prn-p*). Note that this line diverged from the widely used C57BL/6J line many years ago and therefore the two lines are not genetically identical, although limited studies suggest that their response to TSE infection is broadly similar.
- RIII: an inbred strain carrying the a allele of the PrP gene.
- VM: a strain carrying the b allele of the PrP gene, produced at the NPU over 35 years ago, by selective breeding from an outbred colony and maintained as an inbred line since then.
- C57BLxVM: the F₁ PrP heterozygote cross between C57BL and VM.

Health status of mice
Many transmission experiments involve extremely long incubation periods, so it is ethically and economically desirable to reduce the incidence of intercurrent disease in the mouse colony, by maintaining a barrier system to exclude mouse pathogens. The major elements of this are:
1. Restricting staff access to facility, changing clothes and shoes and, ideally, showering on entry.
2. Decontaminating all materials, animal food, bedding etc. entering the facility, for example by autoclaving, irradiation or ethanol spraying.

3. Introducing any new mouse lines into the colony by caesarean derivation or embryo transfer, with appropriate quarantine and pathogen testing procedures.
4. Regular monitoring of the colony for the presence of mouse pathogens.

TSE sources

Strain typing is usually performed using inocula prepared from brain, but other tissues containing appropriate levels of infectivity may also be used. It is extremely important that there should be no opportunity for contamination of the sample with TSE infectivity from another source and that it should be bacterially as clean as possible. The sample should therefore be dissected using aseptic techniques, using sterilised single-use disposable instruments (see comments on equipment below). The sample should be stored frozen in sterilised, single use containers at −20 °C or lower.

Equipment

No specialised equipment or instrumentation is used in these procedures. However, to prevent cross-contamination, any instruments or glassware that come into contact with the infectious sample before injection should be new and disposed of after a single use. This is because some laboratory TSE strains are not completely inactivated by autoclaving, even at 136 °C for 1 h (porous load). Before use, new instruments and glassware should be autoclaved or used directly from suppliers' sterile packs.

Containment

TSEs are categorised as either Hazard Group 2 or Hazard Group 3 organisms, depending on the origin. Transmission studies and the histopathological analysis of the generated tissues should only be undertaken in laboratories and animal facilities with the appropriate level of microbiological containment, as documented in the UK by the Advisory Committee on Dangerous Pathogens (UK Department of Health CJD website).

3 Methods

3.1 General methods

There are three stages of the NPU strain typing approach (Fig. 1):
1. Primary transmission from the naturally infected species to a panel of inbred mouse strains.
2. Secondary passage from individual infected mice in the primary transmission, again into a panel of mouse strains.
3. Continued serial passage in mice, accompanied by testing in a panel of mouse strains.

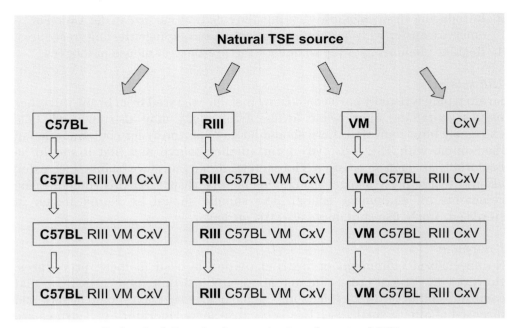

Figure 1   Overall plan for full strain characterization of a natural TSE source.
Serial passage lines are set up in C57BL, RIII and VM (bold), with testing in C57BL, RIII,
VM and C57BLxVM (CxV) at each passage stage.

At each stage, incubation periods and lesion profiles are determined and compared with data generated from a large number of previous transmission and strain typing studies. However, depending on the results in the primary transmission, the type of isolate being characterised and the questions being asked, it may not be necessary to proceed to the second and third stages.

Some isolates may require several serial mouse passages before their incubation periods and lesion profiles become stable. When this has occurred it has been consistent with the gradual selection of variant TSE strains with shorter incubation periods in the new host. As a routine at NPU separate serial passage lines are set up in RIII, C57BL and VM mice, with testing in the full panel at each passage. In practice, the same mouse-passaged strain is usually isolated in RIII and C57BL mice and a different strain is usually isolated in VM mice [14]. It is not known whether these strains were present in the original host, or whether new TSE variants of the original strain(s) have arisen in the course of mouse passage. However, it is clear that the characteristics of the mouse-passaged strains provide useful information about the infection in the original host. For example, the same two mouse-passaged strains have been isolated from all sources of BSE tested [14].

3.2 Primary transmissions

The mouse strain panel used at NPU usually comprises C57BL, RIII and VM inbred strains and the F_1 cross, C57BLxVM. However, depending on the purpose, it may not be necessary to include all four mouse groups in primary transmissions. For example, transmissions are generally difficult to achieve in C57BLxVM mice and these can often be omitted from the primary transmission panel, with minimal loss of useful data. The details of challenge are given in Protocol 1. Because of species barrier effects, the "attack rate" in primary transmissions is often low and incubation periods in the mice that develop disease may be extremely long. To maximise the chances of transmission:

- Mice are inoculated by both the intracerebral and intraperitoneal routes.
- A relatively concentrated (10%) tissue homogenate is inoculated.
- Groups of 20–24 mice of each strain are challenged, to compensate for intercurrent losses.

Concerning the third of these points, the maximum lifespan of an uninfected mouse is about 3 years, but most will die of intercurrent disease or old age considerably earlier. The age at which there is 50% survival varies according to the mouse strain, but is generally around 2 years.

3.3 Secondary and serial passages in mice

Brain samples from individual RIII, C57BL and VM mice from the primary transmission are inoculated into further mouse panels (see Protocol 1). At this stage, because there is no species barrier, transmissions are usually much more readily achieved and incubations are much shorter. Therefore it is possible to use:

- Only intracerebral injection.
- A lower concentration of tissue homogenate (1%).
- Smaller groups of mice (groups of 10–12 are usually adequate).

The protocol is identical for subsequent serial mouse passages.

3.4 Clinical monitoring of mice

The mice are individually identified from ear punches or electronic tags, which relate to individual mouse records. They are monitored routinely for general health status and signs of neurological disease (see Protocol 1). For primary transmissions, a formal TSE clinical scoring system is introduced at 250 days after challenge. This time is chosen because the shortest incubation periods we have seen in primary transmissions to non-transgenic mice have been around 280 days. For secondary passages it is often possible to estimate the likely incubation period, in which case formal clinical scoring starts approximately 50 days before the expected clinical onset.

The clinical signs vary according to the combination of TSE strain and mouse strain, but may include lethargy, hyperactivity, gait alteration, aggression, a pronounced 'startle' reflex, skin scratching and weight loss or gain. Subtle signs are often seen for a few weeks before progressing to definite neurological signs during the last 2–3 weeks of the incubation period. The clinical scoring system (Protocol 1) is designed to be flexible enough determine a clinical end-point based on a range of clinical presentations [6]. The mice are culled at this clinical end-point and tissues are collected for histopathology and further passage. Once the TSE diagnosis has been confirmed histologically, the incubation period for each mouse is calculated as the time in days between challenge and the clinical end-point.

3.5   Histopathological diagnosis and lesion profiling

Histopathological analysis is performed on coded mouse brain sections (Protocol 2). As a routine, TSE diagnosis is based on the observation of characteristic vacuolar degeneration in haematoxylin and eosin (H&E) stained sections, although this may be supplemented by immunohistochemical detection of abnormal accumulations of PrP or reactive astrocytes. It is important to examine brains from all mice histopathologically because: (1) other conditions, particularly in old mice, may be misdiagnosed as TSE disease, and; (2) some successful transmissions produce little or no clinical disease in the mice, but do produce TSE neuropathology. For lesion profiling, the severity of vacuolation is scored from the H&E sections, using low power light microscopy, in nine specified grey matter areas of brain on a scale of 0–5 (Fig. 2, Tab. 1) [3]. Vacuolation is also scored on a scale of 0–3 in three white matter areas (Fig. 2), as some strains produce severe white matter vacuolation and others do not.

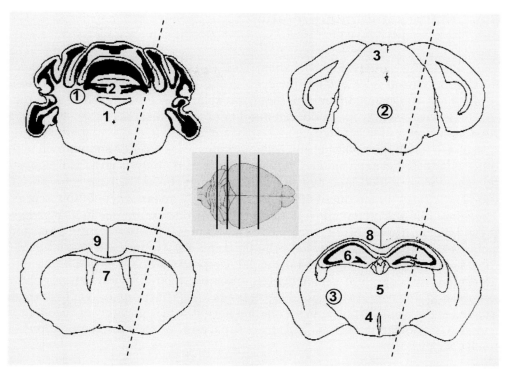

Figure 2 Trimming levels for fixed mouse brain (centre), giving four section levels. The nine grey matter scoring areas (numbers 1–9) and three white matter scoring areas (circled numbers 1–3) are indicated. Grey matter scoring areas are: (1) dorsal medulla, (2) cerebellar cortex of folia adjacent to fourth ventricle, (3) superior colliculus, (4) hypothalamus, (5) medial thalamus, (6) hippocampus, (7) septum, (8) and (9) medial areas of the cerebral cortex at the levels shown. White matter areas are: (1) inferior and middle cerebellar peduncles, (2) decussation of superior cerebellar peduncles, (3) cerebral peduncles. The dashed line shows the lateral portion of brain that is routinely sliced away and frozen for further passage.

Table 1 Definition of vacuolation scores in grey matter brain areas

Vacuolation score	Description
0	No vacuoles
1	A few vacuoles widely and unevenly scattered, not convincingly TSE specific pathology
2	A few vacuoles evenly scattered, more than for score 1 and convincingly diagnostic for TSE
3	Moderate numbers of vacuoles, evenly scattered
4	Many vacuoles with some confluence
5	Dense vacuolation with most of microscopic field confluent, lace-like appearance

3.6   Analysis and interpretation

After decoding and histopathological confirmation of diagnosis, the mean incubation period is calculated for each group of challenged mice and lesion profiles are constructed from the mean vacuolation score in each brain area. Our experience is that the minimum number of mouse brains required to give a reliable lesion profile is 5–6.

A plot of incubation periods and lesion profiles gives a visual comparison with other transmission experiments. However, statistical analysis gives more objective comparisons. For each mouse group in which clinical disease is observed, there will be 1 mean incubation period and 12 mean vacuolation scores. Therefore, for a primary transmission producing disease in four mouse groups there will be a total of 52 numerical data values. If three mouse passaged isolates are established from that primary transmission, this will give a further $52 \times 3 = 156$ values. Cluster and principal components analyses can be used to determine the degree of similarity/difference between different isolates characterised in this way [14]. Depending on the purpose of the analysis, this can be applied to the lesion profile and incubation period data separately or together. It can also be applied to each mouse strain separately or various combinations of mouse strains.

3.7   Use of transgenic mouse lines

A major problem of the use of non-transgenic mice in these studies is that some natural TSE sources either fail to transmit or produce very little clinical disease within the lifespan of the mouse, a phenomenon referred to as the 'species barrier'. A growing number of transgenic mouse lines are available that express PrP from other species (e. g., bovine, ovine, human). The expectation was that transmission to mice would be facilitated by expression of PrP sequences that matched those in the infected donors, in effect removing the species barrier. In practice, this has not always been observed. For example, variant CJD transmits more readily to non-transgenic mice than to mice over-expressing human PrP [13, 15]. The results for a range of transgenic lines challenged with a range of TSE sources have been so inconsistent in this respect that the basic premise, that the species barrier is due to PrP mismatching between donor and recipient, should be re-examined. Nevertheless, some TSEs that are difficult to transmit to non-transgenic mice can be transmitted readily to some transgenic lines and this gives valuable opportunities for expanding the scope of TSE strain typing in mice, by applying similar methods to those described above. However, the strength of the non-transgenic approach is that there is already a very large baseline of data, covering TSEs from a range of different species. The baseline data available for transgenic lines is much more limited.

3.8 Protocols

Protocol 1   Mouse TSE challenge and incubation period determination [6]

1. Homogenise thawed tissue in sterile physiological saline at 10% concentration for primary transmissions or 1% for secondary passages. Then transfer aliquots to sterilised sealable glass containers.
2. If challenges are not to be done immediately, freeze aliquots of homogenate and store at –20 °C or lower.
3. For injection, thaw homogenate and re-suspend by drawing through progressively finer needles, until it is easily drawn up through a 26 g needle.
4. Inject anaesthetised mice using 1 ml syringe fitted with 26 g needle, (a) by both intracerebral (through mid-temporal cortex) and intraperitneal routes for for primary transmissions, or (b) by the intracerebral route only for secondary transmissions. The volumes injected are 0.02 ml intracerebrally and 0.1 ml intraperitoneally.
5. After recovery, return individually coded mice to labelled cages (5–6 per cage).
6. Routinely monitor mice until start of formal TSE clinical scoring (250 days after challenge for primary transmissions, 50 days before expected clinical end-point for secondary passages).
7. Assess mice individually at weekly intervals and give a score of "unaffected", "possibly affected" or "definitely affected".
8. Cull mice when they have received scores of "definitely affected" in two consecutive weeks, or in two out of three consecutive weeks.
9. Maintain mice that do not develop TSE disease for full lifespan or up to a predetermined time point consistent with the aims of the experiment.
10. For collection of tissue for further passage, dissect brain aseptically using new instruments and slice off a lateral third to be frozen in new, sterilised sealable containers (Fig. 2). Fix remaining two thirds of the brain for histological analysis (Protocol 2). If no tissue is required for further passage, fix the whole brain for histology.

Protocol 2   Histopathological procedures, TSE diagnosis and lesion profiling [3]

1. Fix mouse brain if 10% formol saline.
2. Trim coronally at four standard levels (Fig. 2), to give five slices.
3. Process and paraffin wax embed, as shown in Figure 2.
4. Cut 6 μm sections and stain with H&E.
5. Examine by low/medium power microscopy for positive/negative TSE diagnosis based on presence of vacuolation.
6. Score severity of brain vacuolation in 9 grey matter areas on a scale of 0–5 and 3 white matter areas on a scale of 0–3, as shown in Figure 2 and Table 1.

4   Troubleshooting

1. Samples may be bacterially contaminated, particularly those not collected under laboratory conditions. This problem can be minimised by washing the sample in ethanol and rinsing in sterilised saline before homogenisation, or dissecting the sample and discarding original surfaces. Standard bacteriological screening of homogenates before injection will identify any that are still significantly contaminated. Other measures that may be considered are a more prolonged fixation of the sample in ethanol, followed by leaching out with saline before homogenisation, or heating the sample or homogenate to 60 °C for 60 min. Mice may be treated with antibiotic before injection and, in some cases, injected only by the intraperitoneal route. However, most of these procedures result in a slight loss in titre and therefore may prejudice the outcome of already inefficient cross species transmissions.
2. Some samples may contain low titres of infectivity, which will lengthen the incubation period in all mouse groups at the primary transmission stage. In such transmissions, the relative incubation periods in different mouse strains will still give useful information about strain identity. The lesion profile, on the other hand, is not sensitive to infectious dose and can be used for strain characterisation starting with low titre sources.

5   Applications

The most important application of TSE strain typing in mice has been to explore links between novel TSEs in a number of species and BSE in cattle. It has been found that BSE transmits readily to mice, producing clinical disease in close to 100% of challenged animals [11, 12]. When transmitted to the standard mouse strain panel, BSE gives a set of incubation periods and lesion profiles that collectively act as a "signature" for the BSE strain (Figs 3 and 4). Key features are:

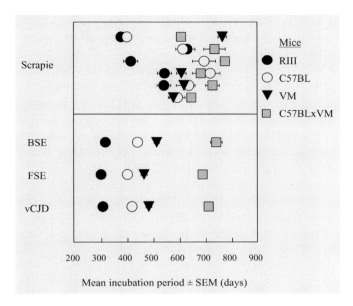

Figure 3 Mean incubation periods in the mouse strain panel in transmissions of BSE, feline spongiform encephalopathy (FSE) and vCJD (pooled data) and scrapie from separate sheep sources

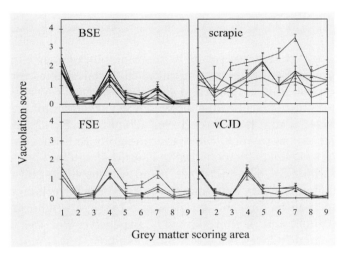

Figure 4 Lesion profiles in RIII mice for individual sources of BSE, sheep scrapie, FSE and vCJD

- Relatively short incubation periods in RIII mice (usually between 300 and 400 days, but sometimes longer for low titre sources).
- Incubation periods longer in C57BL mice than in RIII mice by about 100 days.
- Incubation periods longer in VM mice than in C57BL mice by about 50 days.
- Very long incubation periods in C57BLxVM mice, usually about 700 days.
- Characteristic lesion profiles in all mouse groups.

Furthermore, two mouse-passaged strains have been isolated consistently from BSE sources: 301C isolated in C57BL or RIII mice, 301 V isolated in VM mice [14]. Each of these has its own unique strain signature. The most striking feature is that 301 V has an incubation period of about 110 days in VM mice, approximately 100 days shorter than the incubation period of any other mouse-passaged TSE strain that we have tested.

The above characteristics have never been seen in a series of over 30 sheep scrapie transmissions. There is therefore no evidence from these studies that the BSE strain is present naturally in sheep, although this possibility cannot be ruled out [14]. Unlike BSE, transmissions of sheep scrapie to mice give variable results (e. g., Figs 3 and 4), indicating strain variation in the natural disease, but the extent and significance of this variation are unknown. On the other hand, the BSE signature is seen clearly in transmissions of FSE from cats, novel TSEs from exotic ungulates and, most importantly, vCJD from humans (Figs 3 and 4) [12, 13, 16]. This is the strongest evidence available that these species have been infected accidentally with BSE, probably by dietary exposure. Sporadic CJD has given a unique TSE strain signature that differs from all BSE and scrapie-derived strains [13].

6   Remarks and conclusions

It is clear from the above results that strain typing by transmission to mice can provide evidence of links between TSEs occurring naturally in different species, for example if there is a suspicion that BSE has spread to any other species. A serious concern is that some sheep may have become infected with BSE via contaminated feed, although there is no evidence that this has occurred. The BSE strain signature has been seen in transmissions to mice from experimentally BSE-infected sheep [17]. Therefore this approach has the potential to identify sheep that have been accidentally infected with BSE, but there are some serious limitations. Characterisation of an isolate takes a very long time (up to 2.5 years for primary transmission, plus at least 6 months for each serial mouse passage), requires specialised containment facilities and is therefore very expensive. For these reasons the approach is unsuited to large scale testing of samples, but is more appropriate as a second stage test applied to samples selected by *in vitro* testing.

Acknowledgements

We would like to thank Alan Dickinson and Hugh Fraser who developed this approach and our many colleagues at NPU who have contributed to these studies over the years. We would also like to thank collaborators who provided samples, particularly Bob Will and James Ironside (National CJD Surveillance Unit) and Michael Dawson (formerly at Veterinary Laboratories Agency).

Further reading

Dickinson AG, Meikle VMH (1971) Host-genotype and agent effects in scrapie incubation: change in allelic interaction with different strains of agent. *Mol Gen Genet* 112: 73–79

Fraser H, Dickinson AG (1968) The sequential development of the brain lesions of scrapie in three strains of mice. *J Comp Pathol* 78: 301–311

Bruce ME (2003) TSE strain variation. *Br Med Bull* 66: 99–108

Bruce ME, McConnell I, Fraser H et al (1991) The disease characteristics of different strains of scrapie in *Sinc* congenic mouse lines: implications for the nature of the agent and host control of pathogenesis. *J Gen Virol* 72: 595–603

Bruce ME, Boyle A, Cousens S et al (2002) Strain characterization of natural sheep scrapie and comparison with BSE. *J Gen Virol* 83: 695–704

References

1   Bruce ME (2003) TSE strain variation. *Br Med Bull* 66: 99–108

2   Dickinson AG, Meikle VMH (1971) Host-genotype and agent effects in scrapie incubation: change in allelic interaction with different strains of agent. *Mol Gen Genet* 112: 73–79

3   Fraser H, Dickinson AG (1968) The sequential development of the brain lesions of scrapie in three strains of mice. *J Comp Pathol* 78: 301–311

4   Bruce ME, McConnell I, Fraser H et al (1991) The disease characteristics of different strains of scrapie in *Sinc* congenic mouse lines: implications for the nature of the agent and host control of pathogenesis. *J Gen Virol* 72: 595–603

5   Westaway D, Goodman PA, Mirenda CA et al (1987) Distinct prion proteins in short and long scrapie incubation period mice. *Cell* 51: 651–662

6   Dickinson AG, Meikle VMH, Fraser H (1968) Identification of a gene which controls the incubation period of some strains of scrapie agent in mice. *J Comp Pathol* 78: 293–299

7   Hunter N, Dann JC, Bennett AD et al (1992) Are *Sinc* and the PrP gene congruent? Evidence from PrP gene analysis in *Sinc* congenic mice. *J Gen Virol* 73: 2751–2755

8   Moore RC, Hope J, McBride PA et al (1998) Mice with gene targetted prion protein alterations show that *Prnp*, *Sinc* and *Prni* are congruent. *Nat Genet* 18: 118–125

9   Fraser H (1993) Diversity in the neuropathology of scrapie-like diseases in animals. *Br Med Bull* 49: 792–809

10 Bruce ME, McBride PA, Farquhar CF (1989) Precise targeting of the pathology of the sialoglycoprotein, PrP, and vacuolar degeneration in mouse scrapie. *Neurosci Lett* 102: 1–6

11 Fraser H, Bruce ME, Chree A et al (1992) Transmission of bovine spongiform encephalopathy and scrapie to mice. *J Gen Virol* 73: 1891–1897

12 Bruce M, Chree A, McConnell I et al (1994) Transmission of bovine spongiform encephalopathy and scrapie to mice strain variation and the species barrier. *Phil Trans R Soc Lond B* 343: 405–411

13 Bruce ME, Will RG, Ironside JW et al (1997) Transmissions to mice indicate that 'new variant' CJD is caused by the BSE agent. *Nature* 389: 498–501

14 Bruce ME, Boyle A, Cousens S et al (2002) Strain characterization of natural sheep scrapie and comparison with BSE. *J Gen Virol* 83: 695–704

15 Hill AF, Desbruslais M, Joiner S et al (1997) The same prion strain causes vCJD and BSE. *Nature* 389: 448–450

16 Fraser H, Pearson GR, McConnell I et al (1994) Transmission of feline spongiform encephalopathy to mice. *Vet Rec* 134: 449

17 Foster JD, Bruce M, McConnell I et al (1996) Detection of BSE infectivity in brain and spleen of experimentally infected sheep. *Vet Rec* 138: 546–548

11 Biosafety and Decontamination Procedures

David M. Taylor

Contents

1 Introduction

From the preceding Chapters the reader will already have become aware of the unusual characteristics of the transmissible spongiform encephalopathies (TSEs) and the unconventional nature of their incompletely-characterised causal agents. A notable property of these agents, which will be addressed in this Chapter, is their remarkable degree of resistance to a wide range of chemical and physical inactivation procedures that are effective with conventional microorganisms [1]. As will be discussed, this has a number of implications for the safe handling of TSE-infected materials in the laboratory.

Although the human TSEs do not appear to transmit from one individual to another through normal social contact, they have been transmitted from person-to-person through medical procedures. Creutzfeldt-Jakob disease

Methods and Tools in Biosciences and Medicine
Techniques in Prion Research, ed. by S. Lehmann and J. Grassi
© 2004 Birkhäuser Verlag Basel/Switzerland

(CJD) has been transmitted accidentally from infected to naïve individuals through neurosurgical procedures as a consequence of using inappropriate techniques to sterilise instruments or medical devices that had been in contact with the brain tissue of infected individuals [1]. A survey also indicated that individuals who had been subjected to neurosurgery were at a somewhat higher risk of developing CJD in later years compared with the controls [2]. This might have resulted from the use of inadequate sterilising procedures for the neuro-surgical instruments that were involved but the study did not provide any information regarding the sterilisation procedures that had been used. There has been ongoing concern regarding the potential transmission of CJD from affected to unaffected individuals through medical procedures because of the possible inadequacy of the sterilisation processes that are applied to re-usable medical devices and surgical instruments. As will be discussed, none of the currently recommended autoclaving procedures for the sterilisation of surgical instruments is completely effective with TSE agents under worst-case condi-tions. In addition to the relevance that this might have for the potential iatrogenic transmission of CJD, there are also safety implications for handling TSE agents in the laboratory.

The emergence of variant CJD (vCJD) in the mid 1990s [3], and the sub-sequent occurrence of 146 definite or probable cases in the UK by March 2004, has escalated the general degree of concern regarding the potential person-to-person transmission of CJD-like diseases through the inadequacy of the proce-dures used to sterilise surgical instruments and devices. This enhanced concern is partly due to the fact that it is not possible to predict at present how many more cases of vCJD will occur, but also because many more tissues (especially those of the lymphoreticular system [LRS]) become infected in vCJD-, compared with CJD-infected individuals [4]. Preliminary evidence from the study of vCJD-infected materials from humans, and from the much more wide-ranging studies of scrapie in sheep and laboratory animals, indicate that infectivity is likely to be present in LRS tissues for some considerable time before the central nervous system (CNS) becomes infected, which is when clinical diseases manifests itself. Therefore, surgeons could carry out procedures involving deliberate or inci-dental invasion of LRS tissues in individuals that have no clinical signs of neurological disease but are incubating vCJD. This also means that infected LRS tissues from subclinical cases of vCJD could be sent for investigation to the laboratory following surgery, biopsy or autopsy.

In addition to the cases of CJD that are directly associated with the known failure of sterilisation systems applied to instruments or medical devices [1], a number of cases of iatrogenic CJD has arisen from three other sources. The therapeutic use of hormones derived from cadaveric human pituitary glands has resulted in a significant number of cases of CJD among the recipients [5]. Validation studies carried out on the chromatographic system used most recently in the UK for hormone production showed that this system is capable of inactivating or removing infectivity [6], and a common feature of the UK cases was that they had received treatment between 1974 and 1976 with hormones prepared by procedures that preceded the introduction of the chromatographic method [7]. The use of pituitary-derived growth hormone was discontinued in the UK in 1985, by which time a genetically-engineered alternative was available, but the pituitary-derived product was still used thereafter in other parts of the world. Commercially-processed dura mater obtained from human cadavers is used in surgical repair techniques, and has also caused a significant number of iatrogenic cases of CJD [5]. The only validation study that has been published with regard to the TSE-related safety of commercially-processed dura mater showed that the level of TSE infectivity that might be present was reduced substantially but not completely by a manufacturing process that included exposure to 1 M sodium hydroxide [8]. Initially, it was considered that any CJD infectivity that might be associated with dura mater was likely to be associated with either its natural contact with potentially-infected cerebrospinal fluid or through contamination with infected brain-tissue during its collection. However, later investigations involving the study of dura mater from scrapie-infected mice demonstrated that this tissue does become infected, and that the infectivity levels are only around 100-fold less than those found in the brain tissue of terminally-affected individuals [9]. Another, but much less common, means of transmitting CJD accidentally has been through corneal transplantation [5].

Over the past 50 years, there has been an ever-increasing accumulation of decontamination methods that are effective with conventional microorganisms but not with TSE agents [1]. Although not yet formally demonstrated, it is anticipated that the vCJD agent will share this property because it is the same as the bovine spongiform encephalopathy (BSE) agent [10] which has already been shown to be difficult to inactivate [11]. Even some inactivation procedures that were previously considered to be completely effective with regard to the TSE agents in general are now known to provide a substantial degree of, but not complete, inactivation. Such procedures include exposure to 1 M or 2 M sodium hydroxide for an hour at room temperature [8, 11, 12–14], gravity-displacement autoclaving at 132 °C for an hour [14, 15], porous-load autoclaving at 134–138 °C for 18–60 min [11], or gravity-displacement autoclaving in 5% sodium dodecyl sulphate at 121 °C for 15 min [16]. Nevertheless, the use of sodium hypochlorite solutions containing at least 20,000 ppm of available chlorine still appears to be an effective method although it is not a particularly user- or product-friendly procedure [11, 17]. Although autoclaving or exposure to 2 M sodium hydroxide have not proved to be completely effective *per se* for inactivating TSE agents, inactivation can be achieved by combining these procedures simultaneously or sequentially [18]. Additional data relating to the inactivating capacity of hot hydroxide come from a study in which a particularly thermostable TSE agent was completely inactivated by boiling for 1 min in 1 M sodium hydroxide [19]. Consequently, the use of hot alkali features in four of the six decontamination procedures that have been recommended recently by the World Health Organisation (WHO) [20]; these recommended methods are listed in Table 1, and will be discussed further.

Table 1   Methods for inactivating TSE-infected materials

1. GD autoclaving at 121 °C for 30 min in NaOH[a]. Clean, rinse and subject to routine sterilisation.

2. Immerse in NaOH[a] or NaOCl[b] for 1 h. Rinse. Immerse in water and expose to GD autoclaving at 121 °C for 1h. Subject to routine sterilisation.

3. Immerse in NaOH[a] or NaOCl[b] for 1 h. Rinse, and transfer to an open pan. Expose to PL autoclaving at 134 °C for 1 h. Subject to routine sterilisation.

4. Immerse in NaOH[a] and boil for 10 min. Rinse and subject to routine sterilisation.

5. Immerse in NaOCl[b] (preferred) or NaOH[a] for 1 h at ambient temperature. Rinse and subject to routine sterilisation.

6. PL autoclaving at 134 °C for 18 min.

[a] = 1 M sodium hydroxide solution
[b] = sodium hypochlorite solution containing 20,000 ppm available chlorine

Clearly, working in the laboratory with the BSE agent or TSE agents of human origin requires the use of containment facilities, and level 3 containment is recommended. However, it is recognised that these diseases are unlikely to be transmitted via aerosols, and derogation from the full rigours of category 3 containment is permitted. For example, subject to local risk-assessments, it may not be necessary to a) maintain the laboratory under negative pressure, b) make the laboratory sealable for fumigation (since fumigation is ineffective with TSE agents), or c) HEPA filter extract air. Although numerous studies have failed to establish any association between human disease and dietary or occupational exposure to the scrapie agent [21], it is recommended that this agent should be handled in category 2 containment facilities. Because the data for other animal TSEs are not so extensive, it has been recommended that their causal agents should be handled in category 3 containment facilities [22]. This seems sensible, given that there is now a convincing link between BSE and vCJD [3, 10] that represents the first known transmission of an animal TSE to humans, even though the link is thought to be dietary [23]. There is current concern about chronic wasting disease, another TSE, which was originally confined to elk and mule-deer in the north-western areas of the USA but is now spreading to other regions of the USA, and has spilled over into farmed populations of elk in south-west Canada. While there is no current evidence that this disease transmits to humans, the considerable increase in its incidence is elevating the degree of concern with regard to this possibility [24]. The required standards for containment facilities will not be discussed in any detail because this information is readily available, and the details vary somewhat around the world. However, the UK recommendations are reasonably representative of what is required generally [22].

Safety precautions to protect personnel are obviously appropriate in laboratories that work with, or receive samples that might contain, conventional pathogenic microorganisms. In histopathology laboratories it is generally considered that formalin- or glutaraldehyde-fixed tissues they receive will be largely free from conventional infectious agents because of the well-known disinfecting effect of aldehydes. However, with CJD-like diseases these groundrules change because the causal agents survive fixation in formalin or glutaraldehyde. When hamster brain containing $10^{10.2}$ ID_{50}/g of scrapie infectivity was fixed in formol saline for 48 hours, only 1.5 logs of infectivity were lost [25]. Even after full histological processing, the titre loss was only 2.8 logs [26], and glutaraldehyde fixation is also known to permit the survival of TSE infectivity [27, 28].

Until 1993, CJD had been observed in 24 individuals who had been healthcare workers of various types, including a pathologist and two technicians who had worked in neurohistopathology laboratories, but there was no evident association between their development of CJD and any occupational exposure to the CJD agent [29]. Furthermore, there was an interval of 40 years between the recognition of CJD as a distinct clinical entity and the suspicion that it might be infectious. Although CJD is a rare disease, brain tissue from infected individuals must have been handled worldwide without significant precautions during this period by pathologists and laboratory personnel without any apparent increased incidence of the disease in such individuals. Nevertheless, the accidental transmission of CJD to human recipients of CJD-contaminated human growth hormone by intramuscular injection demonstrates that occupationally acquired disease through trauma is a possibility [5]. This is supported by the data from experiments in which mouse-passaged scrapie infectivity was transmitted efficiently to mice through skin scarification [30]. Consequently, the handling of CJD-infected materials in the laboratory has to be viewed as a procedure that has some element of risk to the personnel involved. To reduce this risk in the histopathology laboratory, one suggestion was to fix tissues in formol saline containing sodium hypochlorite [31]. Although, as already discussed, high concentrations of sodium hypochlorite inactivate TSE agents, there has been no validation of its effectiveness when combined with formalin. The addition of phenol to formol saline has also been suggested as an effective procedure to inactivate TSE agents [32–34] but the basis of these proposals was flawed [35]. Phenolised formalin was shown subsequently to be not only ineffective with regard to inactivation but also resulted in poor fixation [36, 37]. Sections, stained with haematoxylin and eosin, prepared from scrapie-infected formol-fixed brain tissue that had been autoclaved at 134 °C for 18 min retained sufficient integrity to permit the quantitative scoring of spongiform changes by microscopy [38], and it was suggested that autoclaving at 126 °C for

30 min [39] or 132 °C for 6 min [40] could be used to inactivate CJD infectivity in formol-fixed brain-tissue. However, mouse- or hamster-passaged scrapie agent in formol-fixed brain has been shown to survive porous-load autoclaving at 134 °C for 18 min [41] or gravity displacement autoclaving at 134 °C for 30 min [25] with titre losses of less than 2 logs. The only procedure that has been shown to result in significant losses of infectivity in formol-fixed tissues, without any significant loss of microscopic morphology, is a one hour exposure to concentrated formic acid [36]. In that study, the level of infectivity in hamster brain infected with scrapie agent was reduced from $10^{10.2}$ ID_{50}/g to $10^{1.3}ID_{50}$/g. With mouse-brain infected with a CJD agent, the original titre of $10^{8.5}ID_{50}$/g was reduced to $10^{2.3}ID_{50}$/g. However, in another study, where mouse-brain infected with the BSE agent was fixed using paraformaldehyde-lysine-periodate (PLP), a necessary prerequisite for immunocytochemical investigation that is an important aspect of TSE investigation, the degree of inactivation by formic acid was calculated to be 2 logs less than that achieved with formol-fixed scrapie-infected hamster brain, despite the equivalent starting levels of infectivity of the two agents [42]. This suggests either that infected tissues fixed with PLP are less amenable to the inactivating effect of formic acid than those fixed with formalin, or that there is a fundamental difference in the susceptibility of the scrapie agent compared with the BSE agent. The latter is less likely because there is no consistent evidence that TSE agents vary in their degree of susceptibility to inactivation by chemical methods.

There are differences in the actual or perceived risks associated with the handling of TSE agents in the laboratory, depending upon the methods that are involved. For example (as discussed above), any theoretical risks associated with the handling of fixed tissue infected with CJD-like agents can be largely annulled by the treatment of fixed tissue with formic acid. In contrast, the handling of TSE agents in other types of laboratory activities might result, for example, in the disruption of infected tissue by homogenisation that has the potential to release many more infectious airborne particles than section-cutting in the histopathology laboratory, especially if the latter tissues had been treated with formic acid. There is also the capability, through partial purification procedures, of producing samples that contain infectivity titres higher than those found in naturally infected tissues. Apart from general good laboratory practice, the principal recommendation when handling TSE agents under such conditions is to use microbiological safety cabinets, and this will be discussed.

2   Materials

The effectiveness of the chemicals mentioned in the Methods section for the inactivation of TSE agents is probably not dependent upon their degree of purity, and the use of technical grade chemicals is likely to be just as effective as using more costly highly-purified chemicals. Nevertheless, where specific brands of chemicals have been specified in publications where inactivation of TSE agents has been reported, it would seem sensible to use these specific products in everyday practice if possible, but some may no longer be commercially available. Those cited in the studies on successful inactivation using sodium hypochlorite [11, 17] were Sterilex 211 obtained from Brentchem Ltd, Wimbledon, UK, and Kinray obtained from Reddish Savilles Ltd, Cheadle, UK. In the case of the effectiveness of combining treatment with 2 M sodium hydroxide and autoclaving [18], the hydroxide was obtained from BDH Laboratory Supplies, Poole, UK. In the two reports of TSE infectivity levels in fixed tissues being reduced significantly by a formic acid treatment, there was specific information on the formic acid used in only one instance [36]. In this case, the formic acid was ACS grade ($\geq$ 96%), obtained from EM Industries Incorporated, Gibbstown, USA.

With regard to the use of safety-cabinets and autoclaves, as referred to in the Methods section, there are national regulatory standards requiring that such items of equipment must be able to meet minimum performance standards.

3   Methods

3.1   General decontamination and sterilisation

As has been discussed, there are very few methods that reliably inactivate TSE agents under worst-case circumstances. Those that seem to be effective (i. e., strong solutions of sodium hypochlorite or hot sodium hydroxide) are not particularly product- or user-friendly. Consequently, a number of laboratories that handle these types of agents have a policy that involves, as far as possible, single-use, presterilised items that can be incinerated. This policy extends to the disposal by incineration of glass tissue-homogenisers, surgical instruments and other items of equipment that would normally be considered too expensive to discard after a single usage. It should be emphasised that the introduction of this policy is not so much driven by concerns regarding the safety of laboratory personnel, but rather by the wish to avoid cross-contamination events (especially in laboratory animals) that could seriously compromise the outcome of experimental studies. The safety of laboratory personnel should be assured by adopting good and sensible working practices. It is appreciated that not all laboratories in all parts of the world will have access to incineration facilities, or be able to bear the financial burden imposed by disposing of expensive items of equipment. Where the recycling of glassware, instruments etc., is unavoidable, reference should be made to the decontamination and sterilisation methods recommended by the WHO [20] that are listed in Table 1, bearing in mind that these are listed in their order of effectiveness. For example, autoclaving at 134 °C *per se* is the final option on the list but it is known that this procedure may not necessarily be completely effective [11]. Its presence on the list indicates that it would be better to carry out this procedure rather than do nothing if it is not practical to use any of the more effective options. Laboratory discard jars for contaminated pipettes, pipette-tips etc., should contain sodium hypochlorite solutions that have at least 20,000 ppm of available chlorine (see below). Although such solutions have been shown to be completely effective with TSE agents [11, 17], it is a common practice to autoclave discard jars at the end of each day. If this is done, any remaining sodium hypochlorite must be neutralised (e.g., by adding an excess of sodium thiosulphate) to prevent chlorine being released during autoclaving; this can have a deleterious effect on stainless steel autoclave chambers.

3.2   Decontamination of surfaces

Little has been done to determine the efficiency of disinfecting TSE-contami-nated surfaces by simply wiping them with e.g., sodium hypochlorite or sodium hydroxide solutions. All of the testing that has been carried out on the inactivating effects of these compounds has involved the use of fluid suspen-sions of TSE agents. When disinfecting surfaces, it is therefore advisable to keep them moist for an hour with the disinfectant of choice.

3.3   The use of safety cabinets

In using safety cabinets, what must be considered is the resistance of TSE agents to inactivation by formaldehyde that is the customary fumigant for cabinet decontamination. The main objective, therefore, is to adopt working procedures that minimise the potential for contamination. These measures include the use of disposable covering materials on the work surface, and the prevention of aerosol dispersion, e. g., by retaining plugs in tissue-homogeni-sers during sample disruption (and for some time thereafter, if possible). Regardless of these types of precautions, it would be naive to consider that they would guarantee complete freedom from contamination of the internal surfaces of safety cabinets. Although contamination at this sort of level is unlikely to represent any significant risk to the operator, given that such work should always involve the wearing of disposable gowns, gloves and face-masks, the potential for cross-contamination from different TSE sources in laboratory experiments has to be considered. This can be addressed by adopting a routine of washing the internal surfaces of safety cabinets with a solution of sodium hypochlorite containing 20 000 ppm of available chlorine; however, a compromise has to be struck between the perceived necessary frequency of such a decontamination procedure and its potential progressive degradative effect on the surfaces of cabinets. Class II safety cabinets are popular for this type of work because they combine satisfactory degrees of product and personnel protection under conditions that are not too restrictive for the operator. However, the classical design of such cabinets has been such that contamination of the internal plenum and the air-propulsion units is likely. Although this is not problematic for conventional microorganisms that can be inactivated by formaldehyde fumigation that penetrates these areas, there is obviously a problem with TSE agents. Although this type of contamination does not represent any significant risk to the operator or the work activity, there is the problem of how to achieve decontamination before engineers are permitted to carry out repairs or servicing because the plenum and the air-propulsion units are inaccessible as far as hypochlorite decontamination is concerned. There is also the problem of the potential corrosive effects of hypochlorite on the air-propulsion units, even if they were accessible to such treatment. An improvement in this situation was achieved by the manufacture of Class II safety cabinets with the main filters positioned beneath the working surface; this means that contamination of the plenum and the air-propulsion units is avoided unless there is damage to these filters. Because the main filters are readily accessible in such cabinets, it is an easy matter to prevent particle dispersion during their removal, by prior treatment of the filter surface, e. g. with a latex solution or hairspray.

3.4 Formic acid treatment of fixed tissues in the histopathology laboratory

As has been discussed, only small amounts of TSE infectivity are lost when infected tissues are fixed in aldehydes or paraformaldehyde-lysine-periodate (PLP). A substantial increase in the loss of infectivity can be achieved by treating fixed tissues with concentrated formic acid, thus making them safer to handle in the laboratory. This treatment does not significantly impair the microscopic quality of the tissue when examining it subsequently using normal staining procedures or special stains [36]. After fixation, the size to which the tissue is cut before formic acid treatment varies somewhat in different laboratories (see notes in the Troubleshooting section). The cut pieces of tissue are immersed in formic acid (96–100%) for 1 h, after which they are either washed in water or re-immersed in fixative for a short period to remove or considerably dilute the acid before further processing is carried out.

3.5   The use of autoclaves

Although autoclaving standards vary around the world, the methods that have gained prominence with regard to the inactivation of TSE agents are those that have prevailed within the UK and USA. The routine sterilisation of surgical instruments and other items within the UK involves their exposure to porous-load autoclaving for 3 min at a sterilising temperature of 134–137 °C. Although earlier decontamination studies suggested that these standards were probably adequate for TSE agents [17], the UK Department of Health recommended that the sterilising time should be increased to 18 min in circumstances where TSE infection was suspected [43]. Later studies indicated, however, that even this extended recommended time for sterilisation was not entirely effective for TSE agents under worst-case circumstances [11, 44, 45]. In the USA, it is customary to use gravity-displacement autoclaving at 132 °C for 1 h to process surgical instruments and other items. This appeared initially to be effective as far as TSE agents are concerned [46], and became one of the recommended standards for inactivating TSE agents [47]. However, later studies indicated that this process is not completely effective [14, 15].

There are basically two types of autoclave. Gravity-displacement (GD) machines either generate steam by heating water in the bottom of the chamber, or slowly admit steam from an external source into the autoclave chamber. Being lighter than air, the steam migrates to the top of the chamber and, as it builds up progressively, displaces the air downwards. The air is expelled through a thermostatic steam-trap near the bottom of the chamber, and this closes automatically when all of the air has been expelled. This allows the steam-pressure to build up to the required level. The removal of air is an important prelude to the autoclaving process because air-steam mixtures are known to have a poorer sterilising effect than pure steam. GD autoclaves are not suitable for sterilising porous materials like gowns, drapes, towels etc., because air is not efficiently removed from such items. Caution also has to be exercised in the way that items such as instruments are loaded into the autoclave. These should not be placed in any sort of sealed container since the air will not be removed. Nor should they be placed in the bottom of tall, solid-walled receptacles that will trap a lot of air; ideally, they should be laid in shallow perforated trays. Empty, tall containers such as beakers, jugs, measuring cylinders etc., should be laid on their sides so that the air can be removed. The situation with fluids is not so critical since, even if contained in tall, solid-walled receptacles without lids, they will generate their own steam that will displace the air. For fluids such as laboratory reagents, culture-media etc., that are in containers with screw-cap lids, the lids should be sufficiently slack during the autoclaving process to allow air to escape as steam is generated from the fluids within the containers. The lids can be re-tightened immediately after completion of the sterilisation process. It should be noted that some GD machines are equipped with a (usually optional) "drying cycle" in which the steam is allowed to escape rapidly from the chamber after the sterilisation process has been completed. This dries the objects that have been sterilised, and shortens the overall process time. However, this type of cycle is unsuitable for fluids because these would boil off at the end of the process when the chamber is rapidly depressurised.

Porous-load (PL) autoclaves draw a vacuum before and after the sterilising process. The initial vacuum cycle is designed to remove air efficiently from the chamber and from porous items such as gowns, drapes, towels etc. This is further assisted by the second stage of the process in which there are several cycles of rapid pressurisation and depressurisation. The final vacuum cycle, after the sterilisation stage, is designed to remove moisture from the sterilised goods so that they are dry when removed from the autoclave. In contrast to GD autoclaves, PL machines can efficiently sterilise porous fabrics but cannot be used for sterilising fluids because these would boil off under vacuum.

A point for consideration in the use of standard GD or PL autoclaves is that, during the early part of the autoclaving process, infectious (including TSE) agents can be released into the drain-lines and constitute a problem for engineering staff carrying out repairs or maintenance. This can occur through the condensation of steam onto contaminated surfaces, and the condensate being released from the chamber into the drain-lines (as is meant to happen). Alternatively, somewhat larger volumes of infected fluids can be released through the breakage at an early stage of the autoclaving process of glass vessels containing contaminated fluids. Although high-security GD and PL autoclaves are available that retain and sterilise any condensate or spillage, they are not widely used at present. Nevertheless, their use for work with TSE agents or high-risk conventional pathogens should be actively considered, even though this carries a cost penalty.

As has been discussed, none of the routine (or even extended), autoclaving cycles that are currently used to inactivate TSE agents appear to be effective under worst-case conditions. The same is true for exposure to 1 M sodium hydroxide which was previously considered to be an effective method [46, 47] but was later recognised to be incompletely effective [11]. However, the combined or sequential exposure of TSE infectivity to 2 M sodium hydroxide and GD autoclaving at 121 °C appears to be effective [18]. Although sodium hydroxide does not vaporise during autoclaving, precautions have to be taken to avoid it splashing onto the surfaces of the autoclave. This can be achieved by immersing the items to be processed in hydroxide in relatively tall solid-walled containers with a lot of head-room. As discussed above, air entrapment will not be a problem because steam will be generated from the water in the hydroxide solution.

3.6   Disposal of histopathological waste

As has been discussed, TSE agents may not necessarily be inactivated under worst-case circumstances by autoclaving, and their resistance to autoclaving is enhanced considerably by fixation in formalin [25, 41] or ethanol [48]. Although not formally tested, it can be reasonably assumed that this is also likely to apply to glutaraldehyde-fixed tissues. Under such circumstances, up to seven logs of infectivity can survive fixation and autoclaving. Clearly, it is inappropriate to attempt to make TSE-infected histopathological waste safe by autoclaving, and incineration is the customary method of disposal. Nevertheless, experimental studies have shown that TSE-infected formalin-fixed tissues can be made safe by autoclaving in 2 M sodium hydroxide at 121 °C for 30 min [49].

3.7   Disposal of infected liquid waste produced in the laboratory

A variety of laboratory procedures including Western blotting and chromatography produce liquid waste that might be TSE-infected to some extent. There are two reliable methods that can be used to dispose of such fluids. The first is to allow them to be absorbed by sawdust, and then incinerate the sawdust as a solid waste-product. The other option is to collect and pool the potentially contaminated fluids in a storage vessel, and then decontaminate with sodium hypochlorite. To make this a reliable option, the storage vessel should be graduated so that the volume of its contents can be determined. Before the addition of sodium hypochlorite, the pH of the pooled fluids should be checked. If the pH is not between 6 and 8, it should be adjusted to 7 by adding hydrochloric acid or sodium hydroxide as appropriate. By referring to the graduations on the storage vessel, the volume of concentrated sodium hypochlorite solution to be added to achieve a level of 20,000 ppm of available chlorine can be calculated. After a holding period of one hour, any remaining hypochlorite can be neutralised by adding sodium thiosulphate. The precise amount required can be calculated from an iodometric titration. The fluid waste is then safe to release to the sewerage system. Depending upon the amount of organic material that is present in the fluid waste, chlorine could be released during the decontamination process at a level that is detrimental to human health. Therefore, this procedure should only be carried out in a fume-cupboard or a similar safe environment. Clearly, if this procedure was a regular procedure in the laboratory, it could be automated. With automation, the volume of the waste fluids would be registered, and automatically adjusted to pH 7 by the metered addition of acid or alkali. Also, the required amount of hypochlorite for decontamination, and the amount of thiosulphate for neutralisation, could be metered automatically.

4 Troubleshooting

4.1 Sodium hydroxide

It should be noted that 1 M or 2 M sodium hydroxide reacts readily with the carbon dioxide present in air to form carbonates; these neutralise the hydroxide, and diminish its disinfecting capacity. Ideally, working solutions should be prepared fresh for each use, either from solid pellets or by dilution of a 10 M stock solution (this does not react with carbon dioxide). However, 1 M solutions can continue to be used on a day-to-day basis if a pH of 14 is maintained. In principle, sodium hydroxide does not corrode stainless steel but, in practice, some grades can be damaged. It is, however, known to be corrosive to aluminium and glass.

4.2 Sodium hypochlorite

Stock solutions of sodium hypochlorite must be kept sealed because they continuously release chlorine. This process is accelerated by exposure to heat and sunshine, and stock solutions should be stored under cold, dark conditions. Newly-obtained stock solutions generally have a hypochlorite concentration of around 11% but, before preparing working solutions containing 20,000 ppm available chlorine, the precise concentration of the stock solution should be determined using a standard iodometric titration. The dilution factor that is required to prepare working solutions can then be calculated. The slow release of chlorine from stock hypochlorite solutions, even when stored under ideal conditions, means that their concentrations of available chlorine have to be calculated from time to time (usually monthly) to ensure that the dilution factor for preparing working solutions remains appropriate. Because working solutions are much weaker than the stock solutions, and may be used under relatively warm and sunny conditions, they are much less stable. It is therefore recommended that fresh working solutions should be prepared on a daily basis. Sodium hypochlorite corrodes stainless steel but not aluminium or glass. During decontamination, hypochlorite may release sufficient chlorine for this to be a respiratory hazard unless the process is carried out in a well-ventilated or isolated location. In extreme circumstances, respirators may have to be worn.

4.3   Sodium dichloroisocyanurate

The scientific literature suggests that, as far as conventional microorganisms are concerned, the chlorine-releasing compound sodium dichloroisocyanurate (NaDCC) has an equivalent effect to sodium hypochlorite when the solutions have the same available chlorine content [50]. Furthermore, NaDCC is frequently incorporated into pre-weighed commercially-available tablets that are relatively stable and easy to use when preparing working solutions. However, experiments with TSE agents have shown that NaDCC is less effective than hypochlorite at equivalent concentrations of available chlorine [11], and sodium hypochlorite remains the chemical of choice for TSE agent inactivation.

4.4   Formic acid treatment of fixed tissues

It is apparent from consulting various laboratories that there is considerable variation in the size to which fixed tissue is cut before it is treated with formic acid. What should be borne in mind is that in the two studies which demonstrated the inactivating effect of formic acid, the tissue was cut either into slices that were 4–5 mm thick [36], or small pieces that were 8 mm^3 [42]. If the formic acid treatment is used for pieces of tissue that are thicker than 5 mm, it should be determined whether or not the acid actually penetrates into the middle of these pieces, and how long this takes. If necessary, the recommended holding time of 1 h should be increased to allow for the increased penetration time if this is significant.

5   Remarks and conclusions

Clearly, the high degree of resistance of TSE agents to inactivation creates problems regarding their safe handling in the laboratory. Every effort should be made to adopt working practices that avoid or minimise contamination events. It is unfortunate that the only methods that appear currently to be effective with regard to decontamination and sterilisation are harsh procedures that may be harmful to the items being processed, and represent some element of hazard to laboratory personnel. There is clearly a need to develop more benign methods of decontamination that are not harmful to products or personnel.

Further reading

Kimberlin RH, Walker CA, Millson GC et al. (1983) Disinfection studies with two strains of mouse-passaged scrapie agent. *J Neurol Sci* 59: 355–369

Brown P, Rohwer RG, Gajdusek DC (1986) Newer data on the inactivation of scrapie virus or Creutzfeldt-Jakob disease virus in brain tissue. *J Infect Dis* 153:1145–1148

Ernst DR, Race RE (1993) Comparative analysis of scrapie agent inactivation. *J Virol Methods* 41: 193–202

Taylor DM, Fraser H, McConnell I et al (1994) Decontamination studies with the agents of bovine spongiform encephalopathy and scrapie. *Arch Virol* 139: 313–326

Taylor DM (1999) Transmissible degenerative encephalopathies. Inactivation of the causal agents. In: AD Russell, WB Hugo, GAJ Ayliffe (eds): *Principles and Practice of Disinfection Preservation and Sterilisation*. Blackwell Scientific, Oxford, 222–236

Taylor DM. (2000) Inactivation of transmissible degenerative encephalopathy agents: A review. *Vet J* 159: 10–17

References

1   Taylor DM (1999) Transmissible degenerative encephalopathies. Inactivation of the causal agents. In: Russell AD, Hugo WB, Ayliffe GAJ (eds) *Principles and Practice of Disinfection Preservation and Sterilisation*. Blackwell Scientific, Oxford, 222–236

2   Will RG, Matthews WB (1982) Evidence for case-to-case transmission of Creutzfeldt-Jakob disease. *J Neurol Neurosurg Psychiat* 45: 235–238

3   Will RG, Ironside JW Zeidler M et al. (1996) A new variant form of Creutzfeldt-Jakob disease in the UK. *Lancet* 347: 921–925

4   Hill AF, Butterworth RJ, Joiner S et al. (1999) Investigation of variant Creutzfeldt-Jakob disease and other prion diseases with tonsil biopsy samples. *Lancet* 353: 183–189

5   Brown P, Preece MA, Will RG (1992) 'Friendly fire' in medicine: hormones, homografts, and Creutzfeldt-Jakob disease. *Lancet* 340: 24–27

6   Taylor DM, Dickinson AG, Fraser H et al. (1985) Preparation of growth hormone free from contamination with unconventional slow viruses. Lancet ii: 260–262

7   Griffin JP (1991) Transmission of Creutzfeldt-Jakob disease by investigative and therapeutic procedures. *Adverse Drug React Toxicol Rev* 10: 89–98

8   Diringer H, Braig HR (1989) Infectivity of unconventional viruses in dura mater. *Lancet* 1: 439–440

9   Taylor DM, McConnell I (1996) Unconventional transmissible agents in dura mater: significance for iatrogenic Creutzfeldt-Jakob disease. *Neuropathol Appl Neurobiol* 22: 259–260

10 Bruce ME, Will RG, Ironside JW.et al (1997) Transmissions in mice indicate that 'new variant' CJD is caused by the BSE agent. *Nature* 389: 498–501

11 Taylor DM, Fraser H, McConnell I et al. (1994) Decontamination studies with the agents of bovine spongiform encephalopathy and scrapie. *Arch Virol* 139: 313–326

12 Prusiner SB, McKinley MP, Bolton DC et al. (1984) Prions: methods for assay, purification, and characterisation. In: K Maramorosch, H Koprowski (eds.): *Methods in Virology*, Vol. VIII, Academic Press, New York, 293–345

13 Tateishi J, Tashima T, Kitamoto T (1988) Inactivation of the Creutzfeldt-Jakob disease agent. *Ann Neurol* 24: 466

14 Ernst DR, Race RE (1993) Comparative analysis of scrapie agent inactivation. *J Virol Methods* 41: 193–202

15 Horaud F (1993) Safety of medicinal products: summary. *Dev Biol Stand* 80: 207–208

16 Taylor DM, Fernie K, McConnell I et al. (1999) Survival of scrapie agent after exposure to sodium dodecyl sulphate and heat. *Vet Microbiol* 67: 13–16

17 Kimberlin RH, Walker CA, Millson GC et al. (1983) Disinfection studies with two strains of mouse passaged scrapie agent. *J Neurol Sci* 59: 355–369

18 Taylor DM, Fernie K, McConnell I (1997) Inactivation of the 22A strain of scrapie agent by autoclaving in sodium hydroxide. *Vet Microbiol* 58: 87–91

19 Taylor DM, Fernie K, Steele PJ (1999) Boiling in sodium hydroxide inactivates mouse-passaged BSE agent. In: *Abstracts of a Meeting of the Association of Veterinary Teachers and Research Workers*. Scarborough, 29–31 March 1999: 22

20 WHO (1999) WHO/CDS/CSR/APH/2000.3 WHO Infection Control Guidelines for Transmissible Spongiform Encephalopathies: Report of a WHO Consultation. 23–26 March 1999, Geneva.

21 Taylor DM (1989) Bovine spongiform encephalopathy and human health. *Vet Rec* 125: 413–415

22 Advisory Committee on Dangerous Pathogens (1994) *Precautions for Work with Human and Animal Transmissible Spongiform Encephalopathies*. HMSO, London

23 Taylor DM. (2002) Current perspectives on bovine spongiform encephalopathy and variant Creutzfeld-Jakob disease. *Clin Microbiol Infection* 8: 332–339

24 Regalado A (2002) Growing plague of "mad deer" baffles scientists. *Wall Street Journal*, 24 May

25 Brown P, Liberski PP, Wolff A et al. (1990) Resistance of scrapie agent to steam autoclaving after formaldehyde fixation and limited survival after ashing at 360 °C: practical and theoretical implications. *J Infect Dis* 161: 467–472

26 Brown P, Rohwer RG, Green EM et al. (1982) Effects of chemicals, heat and histopathological processing on high-infectivity hamster adapted scrapie virus. *J Infect Dis* 145: 683–687

27 Amyx HL, Gibbs CJ, Kingsbury DT et al. (1981) Some physical and chemical characteristics of a strain of Creutzfeldt-Jakob disease in mice. In: *Abstracts of the Twelfth World Congress of Neurology*, Kyoto, 20–25 September 1981: 255

28 Dickinson AG, Taylor DM (1978) Resistance of scrapie agent to decontamination. *N Engl J Med* 229: 1413–1414

29 Berger JR, David NJ (1993) Creutzfeldt-Jakob disease in a physician: a review of the disorder in health care workers. *Neurol* 43: 205–206

30 Taylor DM, McConnell I, Fraser H (1996) Scrapie infection can be established readily through skin scarification in immunocompetent but not immunodeficient mice. *J Gen Virol* 77: 1595–1599

31 Titford M, Bastian FL (1989) Handling Creutzfeldt-Jakob disease tissues in the laboratory. *J Histotechnol* 12: 214–217

32 Kleinman GM (1980) Case records of the Massachusetts General Hospital (case 45–1980). *N Engl J Med* 303: 1162–1171

33 Brumback RA. (1988) Routine use of phenolised formalin in fixation of autopsy brain tissue to reduce risk of inadvertent transmission of Creutzfeldt-Jakob disease. *N Engl J Med* 319: 654

34 Esiri MM (1989) *Diagnostic Neuropathology*, Blackwell, Oxford

35 Taylor DM (1989) Phenolized formalin may not inactivate Creutzfeldt-Jakob

disease infectivity. *Neuropathol Appl Neurol* 15: 585–586

36 Brown P, Wolff A, Gajdusek DC (1990) A simple and effective method for inactivating virus infectivity in formalin-fixed tissue samples from patients with Creutzfeldt-Jakob disease. *Neurol* 40: 887–890

37 Mackenzie JM, Fellowes W (1990) Phenolized formalin may obscure early histological changes of Creutzfeldt-Jakob disease. *Neuropathol Appl Neurobiol* 16: 255

38 Taylor DM, McBride PA. (1987) Autoclaved, formol-fixed scrapie brain is suitable for histopathological examination but may still be infective. *Acta Neuropathol* 74: 194–196

39 Masters CL, Jacobsen PF, Kakulas BA (1985) Letter to the editor. *Neuropathol Appl Neurobiol* 44: 304–307

40 Masters CL, Jacobsen PF, Kakulas BA (1985) Letter to the editor. *Neuropathol Appl Neurobiol* 45: 760–761

41 Taylor DM, McConnell I. (1988) Autoclaving does not decontaminate formol-fixed scrapie tissues. *Lancet* 1: 1463–1464

42 Taylor DM (1994) Survival of mouse-passaged bovine spongiform encephalopathy agent after exposure to paraformaldehyde-lysine-periodate and formic acid. *Vet Microbiol* 44: 111–112

43 Department of Health and Social Security (1984) *Management of Patients with Spongiform Encephalopathy, Creutz-feldt-Jakob disease (CJD)*. DHSS Circular DA (84) 16. HMSO, London

44 Taylor DM (1999) Inactivation of prions by physical and chemical means. *J Hosp Infect* 43 (Supplement): 569–576

45 Taylor DM (2000) Inactivation of transmissible degenerative encephalopathy agents; A review. *Vet J* 159: 10–17

46 Brown P, Rohwer RG, Gajdusek DC (1986) Newer data on the inactivation of scrapie virus or Creutzfeldt-Jakob disease virus in brain tissue. *J Infect Dis* 153:1145–1148

47 Rosenberg RN, White CL, Brown P et al. (1986) Precautions in handling tissues, fluids, and other contaminated materials from patients with documented or suspected Creutzfeldt-Jakob disease. *Ann Neurol* 19: 75–77

48 Taylor DM (1996) Transmissible subacute spongiform encephalopathies: practical aspects of agent inactivation. In: *Proceedings of the IIIrd International Symposium on Subacute Spongiform Encephalopathies: Prion Diseases*. Paris, 18–20 March 1996: 479–482

49 Taylor DM (2003) Prion inactivation. In: *Abstracts of a Meeting of the National Academies Institute of Medicine Committee on Transmissible Spongiform Encephalopathies; Assessment of Relevant Science*. 21–22 January 2003, Washington

50 Coates D (1985) A comparison of sodium hypochlorite and sodium dichloroisocyanurate products. *J Hosp Infect* 6: 31–40

12 Cell-Free Conversion of Prion Proteins

Louise Kirby and James Hope

Contents

1   Introduction

Prion diseases are progressive neurodegenerative maladies that are characterised by changes in the physical properties and turnover of a glycosyl-phosphatidyl-inositol (GPI)-linked membrane glycoprotein (cellular prion protein, PrP^C). The conversion of the normal, detergent-soluble, proteinase K (PK)-sensitive PrP^C to the abnormal, detergent-insoluble, partially PK-resistant isoform, PrP^{Sc} is a common feature of all these transmissible encephalophies including scrapie in sheep, Creutzfeldt-Jakob disease (CJD) of humans and bovine spongiform encephalopathy (BSE) of cattle [1]. Direct interaction be-

Methods and Tools in Biosciences and Medicine
Techniques in Prion Research, ed. by S. Lehmann and J. Grassi
© 2004 Birkhäuser Verlag Basel/Switzerland

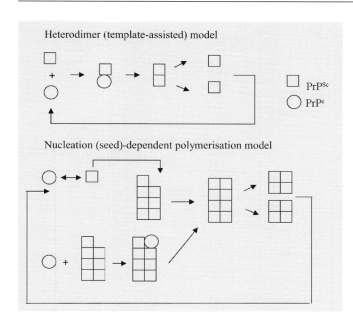

Figure 1 Models of PrP^{Sc} replication.
In the heterodimer (template-assisted) model, PrP^{Sc} exists as a monomer that binds to PrP^{C}, forming a heterodimer and catalysing the conversion of PrP^{C} to PrP^{Sc}. The homodimer than splits to give two PrP^{Sc} seeds for further conversion of PrP^{c}. In the nucleated (seed)-polymerisation model, the conversion of PrP^{C} to PrP^{Sc} is reversible and PrP^{Sc} is stabilised by aggregation.

tween PrP^{C} and PrP^{Sc} is implicated by *in vivo* studies and features in current ideas of molecular pathogenesis and transmissibility such as the heterodimer (template-assisted) and nucleation (seed)-dependent conversion models (Fig. 1). In 1994, Caughey and co-workers introduced a new tool for investigating this process by showing that PrP^{Sc} isolated from the brains of scrapie-affected animals could induce the conversion of radiolabelled recombinant PrP (rPrP) into a PK-resistant isoform, PrP^{res}, *in vitro* [1–5].

Newly formed PrP^{res} is distinguished from the PrP^{Sc} used to seed the conversion by radiolabelling PrP^{C} with ^{35}S cysteine and methionine (Fig. 2). This assay is a rapid and controllable system in which to study the molecules influencing the transition of PrP^{C} to PrP^{Sc} and has been shown to reflect many aspects of transmissible spongiform encephalopathy (TSE) biology such as species specificity, polymorphism barriers and strain properties [6–8]. However, as yet, no *in-*

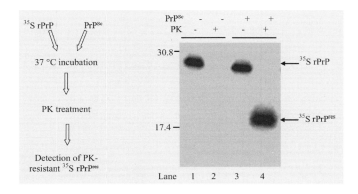

Figure 2 Cell-free conversion reaction.
The left hand flow chart shows the procedure for a cell-free conversion reaction. The right hand panel shows an autoradiograph of a cell-free conversion reaction. Refolded, radiolabelled bacterial mouse recombinant PrP (^{35}S rPrP) was incubated with 87 V scrapie fibrils. ^{35}S rPrP^{res} was detected in the presence (lane 4), but not the absence (lane 2) of PrP^{Sc}.

vitro-generated PrP^res has been shown to be infectious. The cell-free conversion assay has also provided much insight into the kinetics [9, 10] and role of chaperones [11], PrP protein secondary structure [12], di-sulphide bridges [13, 14] and post-translational modifications [15, 16] in the conversion process. It has also been used to gauge the unknown susceptibilities of host species to TSEs [8, 17] and to investigate many lead anti-TSE drugs [18–23].

Traditionally, cell-free conversion assays have used as substrate normal prion proteins purified by immunoprecipitation from mammalian tissue culture cells [5, 24, 25] or by extraction from baculovirus-infected insect cells [26]. Recently, we have introduced bacterial recombinant PrP, refolded *in vitro* into a normal, α-helical, PrP^C-like conformation, as substrate [27]. The use of bacterial recombinant PrP eliminates artefacts that may be due to cellular factors or antibody fragments co-purifying during immunoprecipitation and, additionally, allows structure-activity investigations of protein sequence by site-directed mutagenesis. This methodology, and the replacement of guanidine salts by a more physiological conversion buffer, is presented in this Chapter.

2 Materials

Equipment
- Boiling bath
- Sonicator
- Centrifuges
- SDS-PAGE system
- Phoretix Gel Analysis Software
- Gradifrac™ system
- 37 °C incubator
- 37 °C bacterial shaking incubator
- Dounce homogeniser

Solutions, reagents and buffers
- Solution 1: Methionine free media
- Solution 2: Bacterial lysis buffer, 1 mM ethylene diamine tetra acetic acid (EDTA), 100 mM sodium chloride, 50 mM Tris-HCl pH 8.0.
- Solution 3: IMAC A buffer, 8 M urea, 100 mM sodium phosphate, 10 mM β-mercaptoethanol, 10 mM Tris-HCl pH 8.
- Solution 4: IMAC B buffer, 8 M urea, 100 mM sodium phosphate, 10 mM β-mercaptoethanol, 10 mM Tris-HCl pH 4.5.
- Solution 5: Cation exchange buffer A (IEA), 8 M urea, 50 mM HEPES pH 8.
- Solution 6: Cation exchange buffer B (IEB), 8 M urea, 1.5 M sodium chloride, 50 mM HEPES pH 8.

- Solution 7: Brain lysis buffer, 0.1 M sodium phosphate pH 7.4, 10% (w/v) sarcosine.
- Solution 8: Iodine solution, 0.9 M potassium iodide, 9 mM sodium thiosulphate, 15 mM sodium phosphate pH 8.5, 1.5% (w/v) N-lauryl sarcosinate.
- Solution 9: Sucrose cushion, 20% (w/v) sucrose, 0.6 M potassium iodide, 6 mM sodium thiosulphate, 10 mM sodium phosphate pH 8.5, 1% (w/v) N-lauryl sarcosinate.
- Solution 10: Fixative, 25% (v/v) propan-2-ol, 10% (v/v) acetic acid.
- Solution 11: Conversion buffer, 50 mM potassium chloride, 10 mM magnesium chloride, 100 mM sodium chloride, 50 mM citrate pH 6.5, 0.1% (v/v) Nonidet P-40.

3   Methods

3.1   General methods

Production of radiolabelled recombinant PrP (^{35}S rPrP)

Recombinant PrP (rPrP) of different species has been expressed and purified from a variety of expression plasmids in *Escherichia coli* (*E.coli*). The expression plasmid routinely used in our laboratory is the pTrcHis vector from Invitrogen. It gives high levels of regulated expression of untagged, recombinant proteins from an (IPTG)-inducible promoter (Fig. 3). Figure 4 outlines the production of ^{35}S rPrP.

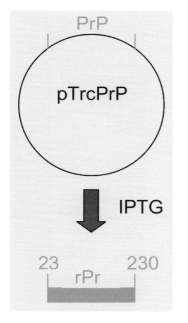

Figure 3 Recombinant PrP (rPrP) expression. Full length mouse rPrP, amino acids 23 to 230, was expressed from the IPTG-inducible pTrc expression vector in *E.coli*. Following growth and expression, rPrP accumulates in inclusion bodies.

Figure 4 The procedure for purification and refolding of rPrP.
Cells are grown, rPrP expression induced and radiolabelled during expression, followed by cell harvest. rPrP is purified by cell lysis, followed by metal affinity and cation exchange chromatography. Following purification, rPrP is refolded *in vitro* by copper oxidation of the disulphide bond and dialysis.

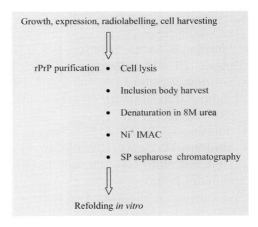

Growth, expression and radiolabelling
1. Cells were grown in 50 ml volumes of methionine free media (Solution 1), 100 µg ml/ml ampicillin, at 37 °C in a shaking incubator at 220 rpm for 4 h.
2. Protein production was induced by the addition of iso-propyl-beta-D-thiogalactopyranoside (IPTG) to a final concentration of 1 mM.
3. 18.5 MBq of Redivue™ L-^{35}S-methionine was added 30 min after induction.
4. Cells were harvested 2 h after induction by centrifugation at 10,000 xg at 10 °C for 15 min.

Cell lysis
1. Cells were resuspended in 1 ml of bacterial lysis buffer (Solution 2), 0.1 M (Phenylmethyl sulfonyl fluoridePMSF), 200 µg/ml lysozyme, per gram of cells and incubated at 4 °C for 20 min.
2. Sodium deoxycholate was added to a concentration of 1 mg/ml and incubated at room temperature (RT) for 30 min.
3. DNase was added to a concentration of 5 µg/ml and incubated at RT for 30 min.

Inclusion body harvest and denaturation
1. Inclusion bodies were harvested by centrifugation at 21,000 × g for 15 min at 10 °C.
2. Inclusion bodies were resuspended in 10 ml of immobilized metal ion affinity chromatography (IMAC) buffer A (Solution 3) per gram of inclusion bodies and incubated at RT for 30 min.
3. Inclusion bodies were harvested by centrifugation at 21,000 × g at 10 °C for 15 min.

Ni+ IMAC
1. 3 ml of Ni^{2+} nitriloacetic acid (NTA) agarose resin (Qiagen) was added to the supernatant and incubated at RT for 1 h with gentle agitation.

2. The resin was applied to a gravity flow column (Qiagen).
3. The column was washed with 10 ml of IMAC buffer A (Solution 3).
4. Resin bound material was eluted with 12 ml of IMAC buffer B (Solution 4).
5. Dithiothreitol (DTT) was added to the eluate to a final concentration of 10 mM.

SP sepharose chromatography
1. A 2 ml SP Sepharose fast flow resin column was equilibrated with IEA (Solution 5).
2. An equal volume of IEA (Solution 5) was added to the Ni-NTA IMAC eluted fraction and loaded onto the column.
3. The column was washed with 20 column volumes of IEA (Solution 5).
4. Proteins were eluted with increasing concentration of IEB (Solution 6).

In vitro *refolding of recombinant PrP*
1. Fractions containing [35]S rPrP were diluted to a concentration no greater than 100 mg/ml.
2. Copper chloride was added in an equal molar ratio and incubated at RT for 3 h.
3. The oxidised protein was dialysed into a 100-fold volume of 50 mM sodium acetate pH 5.5 containing 10 mM EDTA for 4 h.
4. Followed by dialysis into a 100-fold volume of 50 mM sodium acetate pH 5.5, without EDTA, for 4 h. This step was repeated 5 times.
5. [35]S rPrP was concentrated using disposable Microcon 10™ (Amicon).

Optimisation of incorporation of radiolabel into rPrP
To optimise the incorporation of ^{35}S-L-methionine into rPrP, the time after IPTG induction rPrP expression began was determined (Fig. 5). ^{35}S-L-methionine can then be added to coincide with optimal expression of rPrP. Having determined

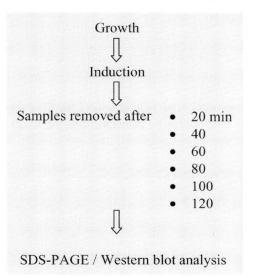

Figure 5   Optimisation of incorporation of radiolabel into rPrP.
The procedure for determining when, after IPTG induction, rPrP expression begins.

Figure 6 Optimisation of incorporation of radiolabel into rPrP.
The procedure for determining the duration of incorporation of ^{35}S-L-methionine into recombinant protein.

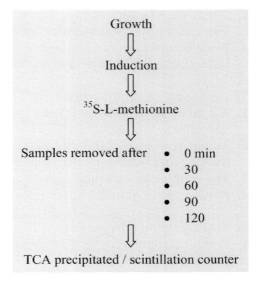

the optimal time for addition of ^{35}S-L-methionine, the duration of ^{35}S-L-methionine incorporation into protein was calculated to determine when to harvest the cells (Fig. 6).

Purification of PrP[Sc]

1. PrP[Sc] was prepared from the brains of terminally-ill scrapie infected animals, essentially as described in Hope et al, 1988 [12, 13] .
2. A 5% (w/v) brain homogenate was prepared in brain lysis buffer (Solution 7) using a 10 ml dounce homogeniser.
3. The homogenate was centrifuged at 22,000 × g at 10 °C for 30 min.
4. The supernatant was centrifuged at 215,000 × g at 10 °C for 150 min.
5. The pellet was resuspended to 3 ml/g of brain with water and incubated at RT for 1 h.
6. The volume of the solution was adjusted to 9 ml/g of brain with iodide solution (Solution 8), layered onto a 3 ml sucrose cushion (Solution 9) and centrifuged at 285,000 × g at 10 °C for 90 min.
7. The pellet was washed in water and centrifuged at 16,000 × g at RT for 30 min.
8. The pellet was resuspended in water, by sonication, to approximately 1 µg/µl.

SDS-PAGE and autoradiography

- Samples were heated in an equal volume of sample loading buffer at 100 °C for 10 min before being loaded into wells and gels were electrophoresed at 180 V.
- Gels were immersed in fixative (Solution 10) for 30 min with gentle agitation before being transferred to a chemical enhancement solution, Amplify (AmershamPharmacia Biotech) for 15 min.
- Gels were dried onto 3 MM filter paper (Whatman) under vacuum at 60 °C for several hours and exposed to film at −80 °C for the appropriate time.

3.2 Methods

Cell-free conversion assay
1. PrP^{Sc} was sonicated and approximately 1 µg incubated with 200 ng of ^{35}S-rPrP in conversion buffer (Solution 11) for 24 h at 37 °C in a 20 µl volume reaction.
2. 20 µl of water was added. A twentieth of the reaction was removed for analysis without PK treatment and the rest treated with 60 µg/ml of PK for 1 h at 37 °C.
3. PK digestion was stopped by addition of Pefabloc™ to 1 mM.
4. Samples were precipitated with 20 µg of bovine serum albumin (BSA) and 4 volumes of ice-cold methanol at –20 °C for 1 h and centrifuged in a bench top centrifuge at 16,000 × g for 15 min.
5. The pellet was analysed by SDS-PAGE and autoradiography.
6. Images were quantified using Phoretix Gel Analysis software.

4   Troubleshooting

- Optimisation of radiolabelling for the chosen expression system
  Different expression vectors and *E.coli* strains can be used to express rPrP. It is important to maximize incorporation of radiolabel into rPrP by choosing an expression vector that shuts down the expression of host proteins once induction of recombinant protein begins.
- Alternative rPrP purification methods
  Purification of rPrP can be enhanced by incorporation of an histidine tag. It may be important to remove the histidine tag following purification to avoid interference with the conversion process.
- Cell-free conversion assays using epitope tagging
  Newly formed PrP^{res} can be distinguished from the PrP^{Sc} used to seed the reaction by the fact that it is epitope tagged rather than radiolabelled and this can be an alternative method of labeling if the use of radioisotopes is restricted in the laboratory. It has been shown, for example, that bacterial, 3F4-epitope tagged rPrP can be converted into PrP^{res} in the cell-free conversion assay. Although epitope tagging avoids the use of radioactivity it limits the range of rPrP variants that can be used as substrate in the cell-free conversion assay and the interpretation of results. Radiolabelling rPrP eliminates such concerns and was therefore chosen as the routine method for labelling rPrP for use in our cell-free conversion assay.

Figure 7 Species specificity in the cell-free conversion assay.
Hamster and mouse ^{35}S rPrP and PrP^{Sc} were incubated in homologous and hetero-logous conversion assays. ^{35}S PrP^{res} was detected in homologous (lanes 2 and 6) but not heterologous (lanes 4 and 8) conversion assays. (Ha) hamster. (Mo) mouse.

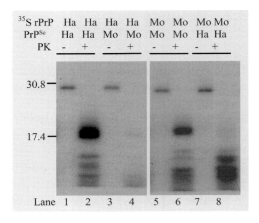

5   Applications

Species barriers
From a public and animal health perspective it is important to know whether particular TSEs can be transmitted between species, particularly with the emergence of novel TSEs in domestic food animals, like BSE in cattle, and the increase in occurrence of some TSEs, such as CWD in deer. The cell-free conversion assay has been used to infer the likelihood of transmission between species. Figure 7 shows how the cell-free conversion assay can be used to model the species barriers of transmission observed in TSE diseases.

Drug screening tool
The advent of variant CJD (vCJD) has provoked a search for effective drug treatments to ameliorate or prevent the effects of prion infection. One of the major approaches to TSE therapy is to modulate PrP^{res} formation. Numerous potential compounds have been identified that inhibit the production of PrP^{res} in cell cultures or in the cell-free conversion assay and some have been shown to prolong the life span of scrapie-infected animals if administered at the time of infection. However, none have been successful in a clinical setting. Figure 8 shows the use of the cell-free conversion assay as a screen for potential drug candidates.

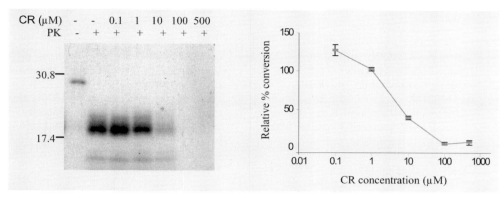

Figure 8 Inhibition of cell-free conversion by Congo Red.
The left hand panel shows an autoradiograph of a cell-free conversion assay with increasing concentration of Congo Red. This is represented graphically in the right hand panel. (CR) Congo Red.

6   Remarks and conclusions

PrP^{Sc} can induce the conversion of PrP^C to a PK resistant form, PrP^{res}, in an *in vitro* cell-free assay. This assay uses as substrate PrP^C immunoprecipitated from mammalian tissue culture cells and has provided extensive information on the conversion of PrP^C to PrP^{Sc}.

This Chapter has described the use of bacterial recombinant PrP, biochemically purified and refolded *in vitro*, as the substrate in the cell-free conversion assay. This quick and simple method has the advantages of producing high yields of prion proteins, free from contaminants or cellular debris, and allows rapid variation of protein sequence using standard cloning and mutagenesis techniques. This provides a means of investigating the conversion of PrP^C to PrP^{Sc} without the need of time-consuming, expensive experiments which use large numbers of animals. Although the role of factors such as dose and route of transmission play an important part *in vivo*, the cell-free assay mimics several fundamental biological properties of *in vivo* TSE transmission. However, it has not yet been possible to generate *de novo* infectious PrP^{res}. This may be due to the ineffective mimicry of some aspect of the process such as the role of the cell membrane or other specific cellular compartment in the conversion pathway. The simplified methodology described here may aid recognition of the "missing link" in the prion saga that turns a dead-end, mis-folded protein into a self-replicating, protein neurotoxin.

References

1   Prusiner SB (1998) Prions. *Proc Natl Acad Sci USA* 95: 13363–13383

2   Caughey B, Kocisko DA, Raymond GJ et al. (1995) Aggregates of scrapie-associated prion protein induce the cell-free conversion of protease-sensitive prion protein to the protease-resistant state. *Chem Biol* 2: 807–817.

3   Caughey B, Raymond GJ, Callahan MA et al. (2001) Interactions and conversions of prion protein isoforms. *Adv Protein Chem* 57: 139–169

4   Kocisko DA, Come JH, Priola SA et al. (1994) Cell-Free Formation of Protease-Resistant Prion Protein. *Nature* 370: 471–474

5   Kocisko DA, Priola SA, Raymond GJ et al. (1995) Species specificity in the cell-free conversion of prion protein to protease-resistant forms: A model for the scrapie species barrier. *Proc Natl Acad Sci USA* 92: 3923–3927

6   Bessen RA, Kocisko DA, Raymond GJ et al. (1995) Non-Genetic Propagation of Strain-Specific Properties of Scrapie Prion Protein. *Nature* 375: 698–700

7   Bossers A, Belt PBGM, Raymond GJ et al. (1997) Scrapie susceptibility-linked polymorphisms modulate the *in vitro* conversion of sheep prion protein to protease-resistant forms. *Proc Natl Acad Sci USA* 94: 4931–4936

8   Raymond GJ, Hope J, Kocisko DA et al. (1997) Molecular assessment of the potential transmissibilities of BSE and scrapie to humans. *Nature* 388: 285–288

9   Horiuchi M, Caughey B (1999) Specific binding of normal prion protein to the scrapie form via a localized domain initiates its conversion to the protease-resistant state. *EMBO J* 18: 3193–3203

10 Horiuchi M, Priola SA, Chabry J et al. (2000) Interactions between heterologous forms of prion protein: Binding, inhibition of conversion, and species barriers. *Proc Nat Acad Sci USA* 97: 5836–5841

11 DebBurman SK, Raymond GJ, Caughey B et al. (1997) Chaperone-supervised conversion of prion protein to its protease-resistant form. *Proc Nat Acad Sci USA* 94: 13938–13943

12 Vorberg I, Chan K, Priola SA (2001) Deletion of beta-strand and alpha-helix secondary structure in normal prion protein inhibits formation of its protease-resistant isoform. *J Virol* 75: 10024–10032

13 Herrmann LM, Caughey B (1998) The importance of the disulfide bond in prion protein conversion. *Neuroreport* 9: 2457–2461

14 Welker E, Raymond LD, Scheraga HA et al. (2002) Intramolecular *versus* intermolecular disulfide bonds in prion proteins. *J Biol Chem* 277: 33477–33481

15 Priola SA, Lawson VA (2001) Glycosylation influences cross-species formation of protease-resistant prion protein. *Embo J* 20: 6692–6699

16 Vorberg I, Priola SA (2002) Molecular basis of scrapie strain glycoform variation. *J Biol Chem* 277: 36775–36781

17 Raymond GJ, Bossers A, Raymond LD et al. (2000) Evidence of a molecular barrier limiting susceptibility of humans, cattle and sheep to chronic wasting disease. *Embo J* 19: 4425–4430

18 Caughey WS, Raymond LD, Horiuchi H et al. (1998) Inhibition of protease-resistant prion protein formation by porphyrins and phthalocyanines. *Proc Nat Acad Sci USA* 95: 12117–12122

19 Chabry J, Caughey B, Chesebro B (1998) Specific inhibition of *in vitro* formation of protease-resistant prion protein by synthetic peptides. *J Biol Chem* 273: 13203–13207

20 Demaimay R, Harper J, Gordon H et al. (1998) Structural aspects of Congo red as an inhibitor of protease-resistant prion protein formation. *J Neurochem* 71: 2534–2541

21 Priola SA, Raines A, Caughey WS (2000) Porphyrin and phthalocyanine antiscrapie compounds. Science 287: 1503–1506

22 Rudyk H, Vasiljevic S, Hennion RM et al. (2000) Screening Congo Red and its ana-

logues for their ability to prevent the formation of PrP-res in scrapie-infected cells. *J Gen Virol* 81: 1155–1164. 2000

23 Demaimay R, Chesebro B, Caughey B (2000) Inhibition of formation of protease-resistant prion protein by Trypan Blue, Sirius Red and other Congo Red analogs. *Arch Virol* 16: 277–283

24 Caughey B, Kocisko DA, Priola SA et al. (1996) Methods for studying prion protein (PrP) metabolism and the formation of protease-resistant PrP in cell culture and cell-free systems. In: HF Baker, RM Ridley (eds): *Prion Diseases.* Humana Press, Totowa, 285–299

25 Kocisko DA, Lansbury PT, Caughey B (1996) Partial unfolding and refolding of scrapie-associated prion protein: Evidence for a critical 16-kDa C-terminal domain. Biochemistry 35: 13434–13442

26 Zhang FP, Zhang J, Zhou W et al. (2002) Expression of PrP[C] as HIS-fusion form in a baculovirus system and conversion of expressed PrP-sen to PrP-res in a cell-free system. Virus Res 87: 145–153

27 Kirby L, Birkett CR, Rudyk H et al. (2003) *In vitro* cell-free conversion of bacterial recombinant PrP to PrP[res] as a model for conversion. *J Gen Virol* 84: 1013–1020

13 Cytotoxicity of PrP Peptides

Roberto Chiesa, Luana Fioriti, Fabrizio Tagliavini, Mario Salmona and Gianluigi Forloni

Contents

Methods and Tools in Biosciences and Medicine
Techniques in Prion Research, ed. by S. Lehmann and J. Grassi
© 2004 Birkhäuser Verlag Basel/Switzerland

1   Introduction

In prion disease, accumulation of PrP^{Sc} and PrP amyloid in the central nervous system (CNS) is accompanied by activation of microglial cells, hypertrophy and proliferation of astrocytes, and degeneration of neurons, leading to variable degrees of atrophy of the target regions. The temporal and anatomical correlation between PrP^{Sc} deposition and the development of neuropathological changes suggests that altered forms of the protein are responsible for the nerve cell degeneration and glial cell reaction [1–5]. In principle, one could directly test the toxic effect of PrP^{Sc} by applying the purified protein to neurons in culture. Although there have been several reports of such experiments [6–8], they are difficult to interpret because of uncertainties about the physical state of the PrP^{Sc}, since detergents needed to keep the protein in solution have to be removed before it can be applied to cell cultures. An alternative strategy is to analyze the effect on cultured neurons of synthetic peptides derived from the PrP sequence.

The concept that misfolding of PrP^C causes a transmissible neurodegenerative disorder has prompted studies to identify the polypeptide segments involved in the conversion. Molecular modeling and nuclear magnetic resonance (NMR) investigations have suggested putative critical regions. Some synthetic PrP fragments from these regions showed both an unusual structural polymorphism and the ability to reproduce the neuropathological hallmark of the disease in a cell culture system [9]. In particular, a synthetic peptide spanning human PrP region 106-126 (PrP106-126) recapitulated several chemicophysical characteristics of PrP^{Sc}, including the propensity to form β-sheet-rich, insoluble, protease-resistant fibrils similar to those found in prion-diseased brains [10, 11]. PrP106-126 also induced apoptotic death of primary neurons [12], proliferation and hypertrophy of astroglia [13], and a remarkable increase in cell membrane microviscosity [14], indicating that this part of the flexible N-terminal region of PrP is likely to be involved in the conformational transformation of the protein and in the pathogenesis.

Although PrP106-126 is not normally found in prion-diseased brains, various N- and C-terminal truncated fragments of PrP are produced in the CNS of sporadic and inherited prion diseases [9, 15–19]. These different types of PrP fragments invariably contain the 106-126 sequence, suggesting that this region may be able to trigger a fundamental pathogenic mechanism common to different forms of prion diseases. The evidence that PrP106-126 activity closely depends on neuronal expression of PrP^C [20–23] also argues for a role of the cellular prion protein in the neurotoxic cascade activated by PrP106-126, and highlights the pathophysiological relevance of this model. The observation that longer synthetic PrP peptides, including PrP82-146, which is identical in sequence to a PrP fragment found in the plaques of Gerstmann-Sträussler-Scheinker (GSS) patients, acquire structural properties reminiscent of the pathogenic isoform, and are toxic to cultured neurons [24–26], further supports the synthetic peptide approach for investigating the molecular basis of prion disorders.

Because short peptides are easy to synthesize and characterize physically, many investigators throughout the world have used PrP106-126 as a prime tool to model prion-induced neurodegeneration. However, because PrP106-126 can adopt different secondary structures in different environments [11, 27], various parameters need to be carefully controlled during synthesis and purification, since these conditions can profoundly affect the final conformation of PrP106-126, and, consequently, its biological activity. In fact, the lack of standardized experimental protocols for peptide preparation and cell culture treatment has sometimes produced variable results in different laboratories, raising concerns about the reliability of the model [28], but see also [29, 30].

This Chapter outlines the procedures currently used in our laboratories for synthesis, purification and analysis of PrP106-126, PrP82-146 and synthetic miniprion (sPrP106). It also reviews the principal experimental contexts in which these peptides are valuable, and discusses some of their advantages. Particular emphasis is laid on how synthetic PrP peptides can be exploited to assay the anti-fibrillogenic activity of compounds, and to develop convenient experimental systems for investigating the cellular and molecular bases of prion pathogenesis.

2   Materials

2.1   Chemicals and solutions

For peptide synthesis, purification and characterization
Fmoc-Ala-OH (4-12-1006), Fmoc-Arg(Pmc)-OH (04-12-1073), Fmoc-Asn(Trt)-OH (04-12-1089), Fmoc-Asp(OtBu)-OH (04-12-1013), Fmoc-Cys(Trt)-OH (04-12-1018), Fmoc-Gln(Trt)-OH (04-12-1090), Fmoc-Glu(OtBu)-OH (04-12-1020), Fmoc-Gly-OH (04-12-1001), Fmoc-His(Trt)-OH (04-12-1065), Fmoc-Ile-OH (04-12-1024), Fmoc-Leu-OH (04-12-1025), Fmoc-Lys(Boc)-OH (04-12-1026), Fmoc-Met-OH (04-12-1003), Fmoc-Ser(tBu)-OH (04-12-1033), Fmoc-Phe-OH (04-12-

1030), Fmoc-Pro-OH (04-12-1031), Fmoc-Thr(tBu)-OH (04-12-1000), Fmoc-Trp(Boc)-OH (04-12-1103), Fmoc-Tyr(tBu)-OH (04-12-1037), Fmoc-Val-OH (04-12-1039), and 2-(1H-benzotriazol-1-yl)-1,1,3,3-tetramethyluranium tetra-fluoroborate (01-62-0015) are from Nova Biochem. Hydroxymethylphenoxy (Wang-type HMP) resin, loading 1.16 mmol/g (400957), and dichloromethane (400142) are from Applied Biosystems. Ethanedithiol (02390), thioanisole (88470) and trifluoroethanol (91690) are from Fluka. α-cyano-4-hydroxycin-namic acid (203072), 2,5-dihydroxybenzoic acid (201346) and sinapinic acid (201345) are from Bruker Daltonics. Acetic anhydride (A6404), acetonitrile (A6914), trifluoroacetic acid (T6,220–0), diethyl ether (30,9996–6), ethylene glycol-bis(2-aminoethylether)-N,N,N′,N′-tetraacetic acid (E4378), formic acid (F0507), Igepal CA-630 (I3021), ninhydrin (N4876), triethylamine (47,128–3), proteinase-K (P2308), phenol (P1037), phenylmethanesulfonyl fluoride (P7626), sodium deoxycholate (D6750) are from Sigma-Aldrich. 1,6-diphenyl-1,3,5-hexatriene (D202) is from Molecular Probes. 4-dodecylaminocarbonyl-fluorene-9-ylmethyl succinimidyl carbonate was kindly supplied by Dr. P. Mascagni.

For cell culture
Basal medium Eagle (41010–026), B27 supplement (17504–036), Dulbecco modified Eagle medium (10938–025), Leibovitz medium (L15) (11415–049), minimal essential medium (22561–021), neurobasal medium (21103–049), penicillin/streptomycin (15140–122) and phosphate-buffered saline (70011–036) are from Gibco BRL-Life Technologies. Fetal bovine serum (FBS) is from Hyclone (CHA111D). Aphidicolin (A0781), cytosine arabinoside (C6645), L-cysteine (C7352), dialyzed FBS (F2442), glutamine (G7513), Hanks balanced salt solution (H9394), HEPES (H0887), 3-(4,5-dimethylthiazol-2-yl)-2,5-diphe-nyl tetrazolium bromide (MTT) (M5655), papain (P3125), phenol red (P3532), poly-D-lysine (P6407), poly-L-lysine (P2636), trypsin (T4665), trypsin inhibitor (T9003) are from Sigma-Aldrich. [Methyl-^{3}H] thymidine is from Amersham (TRA 120). Filter Count is from Packard (6013149). All other reagents are from Sigma-Aldrich.

Solution 1. Cortical neuron dissociating medium: 5.8 mM $MgCl_2$, 0.5 mM $CaCl_2$, 3.2 mM HEPES, 0.2 mM NaOH, 30 mM K_2SO_4, 90 mM Na_2SO_4, 0.5 µg/ml phenol red, pH 7.4; 292 mOsmol.

2.2 Equipment

For peptide synthesis, purification and characterization
Automated Applied Biosystems synthesizer model 433A, semi-preparative C4 column (Symmetry 300, 19 × 150 mm, particle size 7 µm, Waters, Japan), Reflex III™ matrix-assisted laser desorption ionization (MALDI) mass spectro-

meter spectrometer, Jasco J-710 CD spectrometer (Jasco, Easton, MD), program DICROPROT V2.5, Nikon Eclipse E-800, EM109 Zeiss, Oberkoken, Germany electron microscope, image analyzer (Nikon, Japan), microviscosimeter
MV1 (Elscint, Haifa, Israel).

For cell culture
Biosafety tissue culture cabinet; CO_2 incubator; stereoscopic microscope; in-
verted phase contrast microscope; fluorescence microscope; thermostatic incubator equipped with an orbital shaker; spectrophotometer with automatic
microplate reader

3   Methods

3.1   Analysis of PrP peptides

Protocol 1   Peptide synthesis and purification

1. Synthesize PrP peptides by stepwise solid-phase synthesis on an automated
   Applied Biosystems synthesizer model 433A at 0.1 mM scale with hydroxymethylphenoxy (Wang-type HMP) resin from N-(9-fluorenyl)methoxycar-
   bonyl (Fmoc) protected L-amino acid derivatives. Amino acids are activated
   by reaction with 2-(1H-benzotriazol-1-yl)-1,1,3,3-tetramethyluranium tet-
   rafluoroborate. A capping step with acetic anhydride is included after the
   last coupling cycle of each amino acid.
2. Cleave peptides from the resin with a mixture of trifluoroacetic acid (TFA)/
   thioanisole/water/phenol/ethanedithiol 82.5:5:5:5:2.5 (v/v), precipitate peptides with cold diethyl ether, and wash several times with the same solvent
   [11].
3. Purify peptides by reverse-phase (RP)-HPLC on a semi-preparative C4
   column (Symmetry 300, 19 × 150 mm, particle size 7 μm, Waters, Japan)
   with a mobile phase of 0.1% TFA/water (eluent A) and 0.08% TFA/acetoni-
   trile (ACN) (eluent B) using a linear gradient of 0–60% eluent B in 40 min
   with a flow rate of 4 ml/min (for PrP106-126).
4. Collect the fractions containing PrP peptides, lyophilize and store at –80 °C.
5. Verify the identity of peptides by mass spectrometry (MS) (see protocol 4).

Particular care must be taken in the purification of longer peptides such as
PrP82-146 and sPrP106 [25, 26].

6. At the end of the synthesis, when peptides are still attached to the resin (step 1) derivatize the N-terminus with a lipophilic probe (4-dodecylamino-carbonyl-fluorene-9-ylmethyl succinimidyl carbonate) following the method described by Ball and Mascagni [31], with some modifications [25].

7. 1.5 equivalents of the probe (calculated on a hypothetical 100% yield of the synthesis) are dissolved in 0.2 M trifluoroethanol:dichloromethane (DCM) 1:3 (v/v) and added to the polypeptidyl-resin suspended in DCM.

8. Stir the suspension for 2 h and confirm the completion of the reaction by the ninhydrin test (Kaiser test) [32].

9. Evaporate DCM under a gentle stream of nitrogen.

10. Cleave the peptide from the resin with TFA/thioanisole/water/phenol/etha-nedithiol 82.5:5:5:5:2.5 (v/v).

11. Precipitate and wash peptides with diethyl ether. At this point labeled PrP82-146 can be lyophilized and purified according to step 12, while sPrP106 is washed with ACN/water (7:3) and dissolved in 60% formic acid before RP-HPLC purification.

12. Purify labeled peptides by RP-HPLC on a semi-preparative C4 column (Symmetry300, 19 × 150 mm, particle size 7 μm, Waters), with a mobile phase of 0.1% TFA/water (eluent A) and 0.08% TFA/ACN (eluent B) using a linear gradient of 15–50% eluent B in 40 min.

13. Collect the 300 nm-absorbing peak and characterize it by matrix-assisted laser desorption ionization (MALDI) MS.

14. Remove the probe by a 2 h-treatment with 10% triethylamine in ACN/water (1:1) and lyophylize the solution.

15. Dissolve the precipitate with concentrated trifluoroacetic acid (TFA) and precipitate again with diethyl ether.

16. Dissolve the pellet in acetonitrile (ACN)/water (1:1) and lyophilize it. Store lyophilized peptides at −80 °C.

Protocol 2   Prediction of the secondary structure and hydropathic profile

Predicting the hydropathic profile and secondary structure is an essential preliminary step especially when designing PrP peptides containing scrambled segments. For hydropathic profile analysis the program devised by Kyte and Doolittle [33] (http://www.expasy.ch/cgi-bin/protscale.pl) can be applied. The double prediction method [34] (http://npsa-pbil.ibcp.fr/) is used for the secondary structure; to predict the membrane protein topology the TMHMM method, based on a hidden Markov model developed by A. Krogh and E. Sonnhammer, is applied [35] (http://www.cbs.dtu.dk/services/TMHMM/).

Protocol 3   Preparation of peptide stock solutions

1. Dissolve PrP peptides at 5 mg/ml in sterile deionized water or 1 mM sodium acetate, pH 5.5.
2. Determine their solubility by verifying the absence of a visible pellet after centrifugation at 16,000 xg for 10 min.
3. Keep the stock solutions for 2 weeks at −80 °C.

Protocol 4   MALDI mass spectrometry

Fragments or full-length PrP peptides are analyzed with a Reflex III™ MALDI mass spectrometer. A few µl of sample are mixed with an equal volume of a saturated solution of α-cyano-4-hydroxycinnamic acid in ACN/0.1% TFA 1:1 (v:v) (for PrP106-126), 2,5-dihydroxybenzoic acid in ACN/0.1% TFA 1:1 (v:v) (for PrP82-146), or sinapinic acid/0.1% TFA 1:3 (v:v) (for sPrP106), and 1 µl of the mixture is deposited on the MALDI target.

Protocol 5   Circular dichroism spectroscopy

Circular dichroism (CD) spectroscopy is used to identify the secondary structure of the peptides or their assemblies in solution.
1. Spectra are recorded using a Jasco J-710 CD spectrometer (Jasco, Easton, MD) at a scanning speed of 20 nm/min, bandwidth of 2 nm, and step resolution of 0.2–1 nm.
2. Dissolve peptides at a concentration of 0.05–1 mg/ml in deionized water or buffered solution and place the peptide solution in quartz cells with an optical path of 0.1 cm.
3. Incubate peptide at room temperature (RT) for different times (from 1 h to one week) before measurement.
4. Determine the secondary structure content using the program CD Pro package, which includes CDSSTR (singular value decomposition, SVD, in the self-consistent formalism), SELCON3 (self-consisted method) and CON-TIN/LL (ridge regression procedure) [36].

3.2   Fibrillogenic activity

This section outlines the methods used to determine the identity of the PrP regions potentially involved in the conversion from PrP^C to PrP^{Sc}, and to investigate the conditions that influence the formation of PrP amyloid. This experimental approach has been successfully used to study the properties of the amyloid protein purified from GSS patients [26], and can also be applied for the first-level screening of potential anti-prion compounds (see Section 4).

Protocol 6   Sedimentation assay

1. Dissolve peptides at the concentrations of 0.5 and 1 mM in 100 mM Tris-HCl, pH 7.0.
2. Incubate 30 µl aliquots at 37 °C for 1, 4, 8, 24, 48, 72 and 168 h, then chill on ice and centrifuge at 16,000 xg for 10 min at 4 °C.
3. Collect supernatant and dissolve the precipitate in pure formic acid.
4. Analyze supernatants and precipitates by RP-HPLC, expressing peptide concentrations at different times as percentages of the zero-time values [24].

Protocol 7   Assay of protease resistance

1. Dissolve peptides at the concentration of 1 mM in 100 mM Tris-HCl, pH 7.0, containing 1 mM CaCl$_2$ at 37 °C for 24 h (PrP106-126) or 72–96 h (PrP82-146).
2. Incubate peptides with proteinase-K (PK) for 30 min at 1:20 (w/w) enzyme-to-substrate ratio.
3. Stop proteolysis by the addition of ethylene glycol-bis(2-aminoethylether)-N,N,N″,N′-tetraacetic acid at a final concentration of 5 mM.
4. Centrifuge the solution at 16,000 xg for 10 min at 4 °C, and dissolve pellets in 30 µl of 97% formic acid.
5. Analyze 20 µl of the above solution by RP-HPLC.
6. Determine the extent of proteolysis as the percentage of the peptide present in the pellets compared to undigested controls.

The protease-resistant core of long PrP peptides such as sPrP106 can also be analyzed by gel electrophoresis and Western blot:

1. Dissolve sPrP106 at the concentration of 0.2 mg/ml in 10 mM Tris-HCl, pH 7.4/100 mM NaCl, 0.5% Igepal CA-630, 0.5% sodium deoxycholate at 37°C up to 1h.
2. Incubate the peptide with PK for 30 min at 1:20 (w/w) enzyme-to-substrate.
3. Stop the hydrolysis by adding phenylmethanesulfonyl fluoride to a final concentration of 5 mM.
4. Analyze samples by SDS-PAGE and visualize by Coomassie staining or by Western blot.

3.3   Biological activity of PrP peptides

Several studies assessed the effects of synthetic PrP peptides on primary cultures of neurons and astroglial cells, as well as on neuron-like and microglial cell lines [9]. These studies have focused primarily on the neurotoxic and gliotrophic activity of PrP peptides, and on their interaction with cell membranes. Typically, cells are exposed to micromolar concentrations of PrP peptides by diluting the concentrated peptide stock directly into the growth medium, and the biological effects are evaluated after acute or chronic treatment.

This section describes the methods currently used in our laboratories to set up and treat primary cultures of neurons and glia, and the procedures to assay cell viability, proliferation, and cell membrane microviscosity. Detailed protocols for the dissection of brain areas of rodents, and further information on culturing nervous system tissue can be found in [37].

Neurotoxicity

PrP peptides are toxic to neurons cultured from several different brain regions, including the neocortex, hippocampus, and cerebellar cortex. Cortical and hippocampal neurons can be prepared from either rat and mouse fetuses or newborn animals. The latter procedure is particularly advantageous when genetically modified mice are used, since a PCR-based genetic screening can be rapidly done on a tissue biopsy before dissecting the brain tissue.

Protocol 8   Coating tissue culture dishes with polylysine

Primary cultures of neurons and glia are maintained on tissue culture dishes coated with either poly-L- or poly-D-lysine, according to the following procedure:
1. Dissolve poly-L-lysine (0.1 mg/ml) in 1.25% H_3BO_3, 1.91% NaB_4O_7. Dissolve poly-D-lysine (0.05 mg/ml) in H_2O. Filter-sterilize the solution.
4. Coat the plates with enough solution to cover the bottom of the dishes, and incubate either 1 h at 37 °C, or overnight at RT.
5. Remove the coating solution, rinse three times with distilled sterile water, and air-dry in a culture hood. Polylysine-coated dishes can be stored wrapped in aluminum foil at RT up to one month.

Protocol 9   Preparation of primary neurons from rat and mouse embryos

Hippocampal and cortical neurons can be prepared from either rat or mouse fetuses according to the protocol outlined below.
1. Remove brains from fetuses on embryonic day 17 (rat) or 15 (mouse).
2. Dissect out the brain region and collect the tissue in a 15-ml Falcon test tube containing 2 ml of phosphate buffered saline (PBS) supplemented with 100 U/ml penicillin and 100 µg/ml streptomycin (pool two cortices or three hippocampi per tube).
3. Centrifuge 7 min at 100 xg at RT. Discard supernatant and add 2 ml of Dulbecco modified Eagle medium (DMEM) containing 10% fetal bovine serum (FBS) and 2 mM glutamine.
4. Dissociate cells by passing the tissue 10–15 times through a fire-polished Pasteur pipette.
5. Count cells (expect 1–1.5×10^6 cells per dissected brain), and dilute cells to 3×10^5 cells/ml, then plate on culture dishes coated with poly-D-lysine.

6. Culture cells 37 °C in a humidified atmosphere of 5% CO_2, 95% air.
7. After four days in culture, add cytosine arabinoside (10 µM) to inhibit non-neuronal cell division. To further minimize the proliferation of non-neuronal cells, cultures can be maintained in serum-free medium.

Protocol 10   Preparation of cortical neurons from newborn rats and mice

Although originally developed for the preparation of cortical neurons, the following procedure can also be applied for the preparation of hippocampal neurons.

1. Remove brains from two-day-old animals.
2. Dissect out the cerebral cortex and collect the tissue from 4–5 animals in a 15-ml Falcon test tube containing 10 ml of cortical neuron dissociating medium (Solution 1) supplemented with 1 mg/ml L-cysteine and 20 U/ml of papain, and 0.36% glucose.
3. Incubate 30 min at 34 °C.
4. Let the tubes stand at RT until the tissue settles, and discard the supernatant carefully.
5. Add 10 ml of Solution 1 containing 1 mg/ml trypsin inhibitor and 0.36% glucose, and incubate the tubes in a horizontal position for 45 min at RT.
6. Let the tubes stand at RT until the tissue settles, and discard the supernatant carefully.
7. Add 2 ml of minimal essential medium (MEM) containing 10% FBS, 2 mM glutamine and 0.36% glucose.
8. Dissociate cells by passing the tissue 5–6 times through a 2-ml glass pipette, and let the tissue settle.
9. Repeat step 8 for 3–4 times until a homogeneous cell suspension is formed.
10. Centrifuge for 5 min at 200 xg at RT. Discard supernatant, and add 1 ml of MEM containing 10% FBS, 2 mM glutamine and 0.36% glucose.
11. Resuspend the cell pellet very gently using a P1000 Pipetman (Gilson).
12. Count cells (expect approximately 4×10^6 cells per dissected brain), and dilute cells to 2.5×10^5 cells/ml, then plate on culture dishes coated with poly-D-lysine.
13. After 4 h replace the medium with Neurobasal medium supplemented with 0.2% B27.
14. Culture cells 37 °C in a humidified atmosphere of 5% CO_2, 95% air.
15. After two days in culture, add cytosine arabinoside (10 µM) to inhibit non-neuronal cell division. Change medium completely every three days.

Protocol 11   Preparation of cerebellar granule neurons

The following protocol can be used for the preparation of cerebellar granule neurons from either mice or rats.

1. Remove cerebella from 6–7 day-old animals.
2. Dissect out the cerebellar cortex, cut the tissue into 2-mm pieces, and collect the tissue from two animals in a 15-ml Falcon test tube containing 3 ml of L15 supplemented with 100 U/ml penicillin and 100 μg/ml streptomycin.
3. Triturate the tissue using a fire-polished Pasteur pipette.
4. Centrifuge for 1 min at 200 xg at RT. Discard supernatant, and add 5 ml of Hanks balanced salt solution containing 0.3 mg/ml trypsin.
5. Incubate 15 min at 37 °C.
6. Add 5 ml of BME containing 10% dialyzed FBS, 25 mM KCl, and 0.5 mg/ml trypsin inhibitor.
7. Centrifuge for 1 min at 200 xg at RT. Discard supernatant, add 5 ml of BME containing 10% dialyzed FBS, 25 mM KCl, and 0.5 mg/ml trypsin inhibitor.
8. Dissociate cells by passing the tissue 10 times through a fire-polished Pasteur pipette.
9. Let the tissue settle and transfer the supernatant containing the dissociated cells into a clean 15-ml Falcon tube. Repeat this step three times and pool the dissociated cells.
10. Centrifuge for 6 min at 600 xg at RT. Discard supernatant, and resuspend the cell pellet in 4 ml of BME containing 10% dialyzed FBS, 25 mM KCl, using a fire-polished Pasteur pipette.
11. Count cells (expect approximately 4–5×10^6 cells per cerebellum), dilute cells to 3.5×10^5 cells/ml, then plate on culture dishes coated with poly-L-lysine.
12. Culture cells at 37 °C in a humidified atmosphere of 5% CO_2, 95% air.
13. After 24 h (rat) or 36 h (mouse) add 3.3.μg/ml aphidicolin to inhibit non-neuronal cell division. Change medium completely every two days.

Gliotrophic activity
The effects of synthetic PrP peptides on astrocytes and microglia are investigated using primary cultures of either mouse or rat glia. The role of glial activation in the neurotoxicity of synthetic PrP fragments can be explored using co-cultures of glia and neurons.

Protocol 12 Preparation of primary astroglia and microglia

The following protocol can be applied for the preparation of glial cells from either mice or rats.

1. Remove brain from 0–4 day-old animals.
2. Dissect out the cerebral cortex, and cut the tissue into 2–5-mm pieces.
3. Collect the tissue from one animal in a 15-ml Falcon test tube containing PBS supplemented with 100 U/ml penicillin and 100 µg/ml streptomycin.
4. Centrifuge for 1 min at 200 xg at RT. Discard supernatant, and add 2.5 ml of DMEM containing 10% FBS and 2 mM glutamine.
5. Dissociate cells by passing the tissue through a fire-polished Pasteur pipette.
6. Pool the cell suspensions, and plate the equivalent of one rat (or two mouse) brains in a 25 cm² flask (Falcon) in a final volume of 5 ml.
7. Culture cells at 37 °C in a humidified atmosphere of 5% CO_2, 95% air.
8. Change half the medium every two days. Let cells grow to confluence (approximately 2–3 weeks).
9. Replace medium with 3 ml of fresh DMEM containing 10% FBS and 2 mM glutamine, and shake flasks overnight in an orbital shaker at 150 rpm at 37 °C.
10. Collect the medium containing the floating microglia into 15-ml Falcon tubes (transfer the medium immediately to prevent cells settling again). Go to step 13.
11. Rinse the adherent astroglia with PBS, and add 0.5 ml of PBS containing 0.05% trypsin.
12. Incubate at 37 °C until cells detach from the substrate (2–5 min), and resuspend the cells in 5 ml DMEM containing 10% FBS and 2 mM glutamine.
13. Count cells (expect approximately 1 × 10⁶ microglial cells per flask, or 5 × 10⁶ astroglial cells per flask), dilute cells to 5 × 10⁵ cells/ml, then plate on culture dishes coated with poly-D-lysine.
14. Culture both astroglia and microglia at 37 °C in a humidified atmosphere of 5% CO_2, 95% air, and replace half the medium every two days.

Protocol 13 Co-cultures of neurons and glia

1. Plate neurons prepared according to Protocols 9–11 on polylysine-coated 96-well plates.
2. Prepare astrocytes or microglia as described in Protocol 12, and plate cells on polylysine-coated Nunc Tissue Culture Inserts (Nunc, 136730).
3. The next day transfer the inserts into the 96-well plate containing the neurons.

Culture treatment and evaluation of the cellular responses
For cell culture treatment, lyophilized peptides are dissolved directly in sterile
H_2O at a final concentration of 2 mM (PrP106-126), 1 mM (PrP82-146) and 0.5
mM (sPrP106). Cultures are exposed either acutely or chronically to PrP
peptides at concentrations between 0.5 and 100 µM. For short-term treatment,
peptides are applied once at the time of plating or after seven days in culture;
for long-term treatment, the peptides are added to the culture medium the day
after plating, and the treatment is repeated every second day. Control cultures
are exposed to the scrambled peptides or to the vehicle only.

Protocol 14   LDH assay

This is a commonly used method for the quantification of cell death, based on
measuring lactate dehydrogenase (LDH) activity released from the cytosol of
damaged cells into the culture medium. Several commercial kits are available
that are convenient and easy to use. We have been successfully using the
"Cytotoxicity Detection Kit (LDH)" by Roche (1644793), following the manufacturer's instructions.

Protocol 15   MTT assay

This method measures the level of cellular reduction of 3-(4,5-dimethylthiazol-
2-yl)-2,5-diphenyl tetrazolium bromide (MTT) to formazan [38], and can be
employed to evaluate either the neurotoxic or proliferative effect of PrP peptides.
1. Prepare a solution of 4 mg/ml MTT in PBS.
2. Add MTT solution to the cell culture medium (1:10 volume, final concentration 0.4 mg/ml), and incubate for 3 h at 37°C in a CO_2 incubator.
3. Remove culture medium carefully and dissolve cells in 0.04 N HCl in 2-
   propanol.
4. Analyze spectrophotometrically at 540 nm with an automatic microplate
   reader.

Protocol 16   Staining with crystal violet

1. Prepare a solution of 0.5% crystal violet in $H_2O:CH_3OH$ 4:1 and filter the
   solution through a paper filter.
2. Remove culture medium carefully and wash cells with PBS. Add enough
   crystal violet solution to cover the cell monolayer. Incubate for 5 min at RT.
3. Remove the staining solution (can be saved and reused 2–3 times), and
   wash the plates thoroughly in tap water.
4. Let the plates dry, and dissolve the cells in 0.1 M Na citrate: 95% ethanol 1:1.
5. Analyze spectrophotometrically at 570 nm with an automatic microplate
   reader.

Protocol 17 Assay of glial cell growth rate by [methyl-³H] thymidine incorporation

This method is used to assess the growth rate of glial cells exposed either chronically or acutely to PrP peptides. To enhance the proliferative effect of short-term treatment, the cultures can be shifted to serum-free medium 24 h before exposure to the peptide [39].

1. Plate cells in 96-well plates as described in Protocol 12.
2. Add 2 µCi/ml [methyl-³H] thymidine to the culture medium and incubate at 37 °C for 24 h.
3. Remove medium and wash cell thoroughly with PBS.
4. Lyse cells with 0.2 ml/well of distilled water, dilute cell lysate in Filter Count, and count samples in a beta-scintillator (expect approximately 2000 cpm for the untreated samples).

Protocol 18 Determination of cell membrane microviscosity

Membrane microviscosity is assessed in suspensions of primary cultures of rat cortical neurons using 1,6-diphenyl-1,3,5-hexatriene (DPH) as a fluorescent probe, as previously described [14]. The reported fluorescence polarization value (FP, expressed as arbitrary units) is a function of the emission (420 nm) detected through an analyzer oriented parallel (p_1) and perpendicular (p_2), to the direction of polarization of the exciting light (365 nm), according to the equation $FP = (p_2-p_1)/(p_2+p_1)$. Membrane microviscosity (η, poise) is related to FP according to the equation $\eta = 2FP/0.46\text{-}FP$.

1. Mechanically detach neuronal cells (2×10^6 cells/ml) and centrifuge the cell suspension at $550 \times g$ for 10 min.
2. Wash cells with saline, resuspend in 2.5 ml of 5 mM phosphate buffer, pH 7.4, containing 2 µM DPH, and incubate for 30 min at RT.
3. Determine the FP value at 25 °C, before and 30 min after the addition of 25 µM of PrP peptides to cell suspensions.

4 Troubleshooting – What to do if PrP peptides are poorly active

Scientists working with amyloidogenic peptides are used to some degree of variability in the biological activity of different batches. Differences in the self-aggregation properties of independent peptide preparations may be at the basis of this variability. It is commonly assumed that assembly of amyloidogenic peptides into fibrils is a prerequisite for peptide toxicity. However, our own experience with PrP106-126 indicates that neurotoxicity is highest when peptide monomers are in equilibrium with polymerized fibrils. Fibril formation is a stepwise process during which small peptide oligomers and protofibrils are formed, and then assemble into amyloid. These intermediate molecular species may be the actual neurotoxic species [40–43]. Therefore, even small differences in the procedure of peptide synthesis and purification may affect the kinetics of fibril formation, and result in significant differences in the biological activity of the peptide.

When a peptide is not toxic, even though it is correct in sequence and pure, it is usually sufficient to dissolve it in concentrated TFA and precipitate it with cold diethyl ether. The precipitate is then dissolved in ACN/water 1:1 and lyophilized, and this step is repeated several times. ACN favors the formation of the β-sheet secondary structure that is the main determinant of peptide fibrillogenicity [44]. The same procedure can be used when an originally toxic peptide preparation becomes inactive after several months of storage at –20 or –80 °C.

5   Applications

5.1   Investigating the molecular mechanism of the conformational conversion of PrP^C into PrP^{Sc}

The structure of both PrP isoforms must be known to understand the molecular mechanism of the conformational transition from PrP^C to PrP^{Sc}. Unfortunately, the insolubility, heterogeneity and complexity of PrP^{Sc} preparations have hampered structural studies of the pathogenic isoform. Synthetic peptides and recombinant fragments of PrP have been more useful for such investigations. Synthetic peptides offer advantages over recombinant proteins: they are homogeneous and free of cellular contaminants. Homogeneity of PrP peptides is of particular importance because potential cofactors in the process of PrP^C→PrP^{Sc} conversion do not complicate the experiment [45]. Moreover, synthetic PrP peptides can be easily modified in specific sites of their primary sequence to facilitate structural and functional studies. Recent studies on the structural and self-aggregation properties of synthetic miniprion, and peptides homologous to sequence 82-146 of human PrP exemplify this approach [25, 26].

Investigations of the structural and amyloidogenic properties of PrP82-146, and of PrP82-146 peptides in which the amino acid stretches 106-126 or 127-146 had been scrambled, indicated that the ability of PrP82-146 to assemble into fibrillar β-sheet structures depended on the integrity of its C-terminal region, and was driven by electrostatic interaction between Asp/Glu residues at the C-terminus, and Lys residues in the N-terminus. Synthetic peptides can also be exploited to investigate the effect of pathogenic mutations on the structural properties and biological activity of PrP [46].

5.2 Testing anti-prion drugs

The need for therapies for prion diseases has become even more pressing with the emergence of the new variant CJD (vCJD). Because PrP[Sc] is critical for disease transmissibility and pathogenesis, it is a primary target for therapeutic intervention. One possible approach is to develop compounds that bind selectively to PrP[Sc] and can destabilize the misfolded structure of the protein, thus facilitating its clearance. Synthetic PrP amyloid and large PrP peptides, that can be folded into different conformers, can be used to model PrP[Sc] for screening compounds *in vitro*. The methods outlined in section 3.2 can be easily applied to test the ability of compounds to bind PrP amyloid and reverse the scrapie-like properties of synthetic PrP peptides. When the drug under examination is fluorescent the binding to PrP peptides can easily be evaluated by fluorescence microscopy [47], and the drug-peptide interaction can be determined by fluorimetric analysis [24]. Thioflavine binding and HPLC quantification of the peptide, with electron microscopy determination, are the methods used to evaluate antifibrillogenic and de-fibrillogenic capacity [24]. The anti-amyloidogenic effect is determined by dissolving the drugs and the peptide simultaneously, while the de-fibrillogenic capacity is tested by adding the substances to the peptide preparation once the fibrils are already formed, usually after incubation at 37 °C for 48–96 h. The aggregated and monomeric peptides are then separated by centrifugation and quantified. The anti-amyloidogenic effect is associated with a reduction of peptide resistance to protease digestion. At the end of the co-incubation with the test substances, PrP peptides are subjected to PK digestion and the residual peptide is determined by HPLC.

These biochemical assays can be combined with experiments aimed at evaluating the extent to which the drugs reduce the neurotoxicity of the peptides. The peptides and the substances to evaluate can be added directly to the culture medium, or administered to the cultures after co-incubation. The latter is recommended when the fibrillogenic rate of the peptide is slow, and fibril formation requires a high concentration (e. g., PrP82-146). Conversely, if the peptide assembles readily into fibrils (e. g., PrP106-126) the co-incubation step is unnecessary.

Although it was postulated that the toxicity of amyloidogenic peptides was associated with their propensity to assemble into fibrils, we found that the toxicity of PrP106-126 was not strictly dependent on its amyloid state [48]. A number of studies suggest that protofibrils or small oligomers, rather than highly polymerized fibrils, are the primary toxic species of β-amyloid and PrP [40–42, 49]. A recent study showed that an antibody raised against a conformational epitope common to soluble oligomers formed by several different peptides, including PrP106-126, can inhibit their neurotoxic effect [43]. Thus, it is likely that drugs that reduce the toxicity of PrP106-126 interact with peptide monomers or small oligomers rather than with large peptide fibrils.

Once the effectiveness of the candidate drug has been evaluated with the methods outlined above, its ability to reverse the abnormal properties of partially purified PrP^{Sc} and decrease prion infectivity can be tested *in vitro* and *in vivo*. Following this strategy, the efficacy of several potential anti-prion compounds has been evaluated [50, 51].

5.3   Investigating the interaction of PrP peptides with cell membranes

Plasma membranes appear to be central to the pathogenesis of prion diseases. The cell membrane is the putative site of interaction between PrP^{Sc} and PrP^{C}, and key events in the conversion process may take place at either the plasma membrane itself or in an endocytic pathway. The detergent-resistant subdomains of the plasma membrane (rafts) may also be involved in the acquisition of protease resistance of PrP in scrapie-infected cells, and in cells expressing PrP molecules carrying pathogenic mutations [52].

A trasmembrane form of PrP that inserts into the lipid bilayer between aminoacids 112 and 135 has been proposed as a neurotoxic intermediate in both genetic and infectious prion disorders [53]. Synthetic peptides containing the transmembrane domain of PrP interact with artificial and natural membranes, and markedly increase their lipid density and microviscosity [14, 25, 26]. These membrane modifications may cause receptor and channel dysfunction of nerve and glial cells, and may account for the cell responses to the peptide *in vitro* [54, 55]. Steady-state fluorescence anisotropy to determine changes in membrane microviscosity is a convenient way to evaluate the ability of PrP peptides to interact with cell membranes.

5.4   Studying the effects of PrP peptides on neuronal and glial cultures

Many studies have investigated the cellular alterations caused by PrP106-126, and suggested potential neurotoxic mechanisms. However, the precise molecular pathways activated by the peptide and the role played by PrP^{C} in mediating PrP106-126 toxicity are not completely understood. An alteration of cellular ion homeostasis, or signaling pathways, possibly through an interaction with PrP^{C}, may account for some of the effects of PrP106-126 [9, 54].

It has been argued that the neurotoxicity of PrP106-126 is strictly dependent on activation of the microglia cells [56]. However, PrP106-126 has several biochemical effects on neurons themselves, which may account for its toxicity. In fact, it is toxic to glia-free neuronal cell lines such as PC12, and mouse and human neuroblastoma cells, as well as endothelial and HeLa cells [57–62]. This evidence does not rule out that microglia activation may contribute to the neurotoxicity of PrP106-126, but indicates that microglia is not an obligate mediator of the toxicity.

Using the methods described in section 3.3 the biological effects of PrP peptides, including peptides whose native primary structure has been modified [27, 46, 63], can be easily explored. Unlike the situation for β amyloid in Alzheimer's disease, PrP106-126 is not actually a fragment found in the brains of humans or animals with prion diseases. Thus, studies with longer synthetic peptides corresponding to PrP fragments actually found in prion-diseased brains are especially valuable [26].

6 Remarks and conclusions

Since the neurotoxicity of PrP106-126 was first reported in 1993 [12], many different laboratories throughout the world have used this peptide to investigate the biological activity of PrP^{Sc}. The evidence that PrP knockout neurons were resistant to the toxicity of PrP106-126 highlighted the pathophysiological importance of this approach, and suggested that the peptide might kill neurons by altering the properties of cellular PrP. However, current data indicate that PrP106-126 does not induce conversion of PrP^{C} into PrP^{Sc}, and suggest that it may recapitulate the toxic, but not the infectious properties of the pathogenic isoform [64]. In fact, several alternative molecular forms of PrP have neurotoxic properties in animal or cell models [49, 65, 66], indicating that PrP^{Sc} may not be the only culprit [67]. It is an important challenge to elucidate the physical structure of these toxic PrP molecules, and establish whether they are intermediates or by-products of prion replication.

Synthetic PrP peptides with toxic properties may prove useful to elucidate the structural and chemicophysical features underlying PrP neurotoxicity, and to define specific mechanisms that contribute to cellular damage. Ultimately, identifying the toxic forms of PrP will be essential for designing therapies for prion diseases.

Acknowledgements

We are grateful to Nadia Angeretti and Ilaria Bertani, who helped to develop the cell culture procedures described in this Chapter, and to Valentina Bonetto for a critical reading of the manuscript. Financial support was obtained from Telethon-Italy (S00083), the Italian Ministry of Health and the European Community (QLG-CT-2001-2353, QLRT-2001-00283). L.F. was supported by fellowships from the Fondazione Monzino. R.C. is an Assistant Telethon Scientist (DTI, Fondazione Telethon).

Further reading

Caughey B (ed) (2001) Prion proteins. In: FM Richards, DS Eisenberg (eds): *Advances in Protein Chemistry. Volume 57*. Academic Press, San Diego.
Wetzel R (ed) (1999) *Methods in Enzymology, Volume 309: Amyloids, Prions and Other Protein Aggregates*. Academic Press, San Diego
Anonymous (1992) Special issue: Is beta-amyloid neurotoxic? *Neurobiol Aging* 13: 535–625
Chiti F, Stefani M, Taddei N et al. (2003) Rationalization of the effects of mutations on peptide and protein aggregation rates. *Nature* 424: 805–808

References

1   DeArmond SJ, Mobley WC, DeMott DL et al. (1998) Changes in the localization of brain prion proteins during scrapie infection. *Neurology* 50: 1271–1280

2   Jendroska K, Heinzel FP, Torchia M et al. (1991) Proteinase-resistant prion protein accumulation in Syrian hamster brain correlates with regional pathology and scrapie infectivity. *Neurology* 41: 1482–1490

3   Williams A, Lucassen PJ, Ritchie D, Bruce M (1997) PrP deposition, microglial activation, and neuronal apoptosis in murine scrapie. *Exp Neurol* 144: 433–438

4   Bruce ME, McBride PA, Farquhar CF (1989) Precise targeting of the pathology of the sialoglycoprotein, PrP, and vacuolar degeneration in mouse scrapie. *Neurosci Lett* 102: 1–6

5   Jeffrey M, Martin S, Barr J et al. (2001) Onset of accumulation of PrP[res] in murine Me7 scrapie in relation to pathological and PrP immunohistochemical changes. *J Comp Pathol* 124: 20–28

6   Muller WE, Ushijima H, Schroder HC et al. (1993) Cytoprotective effect of NMDA receptor antagonists on prion protein (prion[sc])-induced toxicity in rat cortical cell cultures. *Eur J Pharmacol* 246: 261–267

7   Giese A, Brown DR, Groschup MH et al. (1998) Role of microglia in neuronal cell death in prion disease. *Brain Pathol* 8: 449–457

8   Post K, Brown DR, Groschup M et al. (2000) Neurotoxicity but not infectivity of prion proteins can be induced reversibly *in vitro*. *Arch Virol Suppl* 16: 265–273

9   Tagliavini F, Forloni G, D'Ursi P et al. (2001) Studies on peptide fragments of prion proteins. *Adv Protein Chem* 57: 171–201

10 Selvaggini C, De Gioia L, Cantu L et al. (1993) Molecular characteristics of a protease-resistant, amyloidogenic and neurotoxic peptide homologous to residues 106–126 of the prion protein. *Biochem Biophys Res Commun* 194: 1380–1386

11 De Gioia L, Selvaggini C, Ghibaudi E et al. (1994) Conformational polymorphism of the amyloidogenic and neurotoxic pep-

tide homologous to residues 106–126 of the prion protein *J Biol Chem* 269: 7859–7862

12 Forloni G, Angeretti N, Chiesa R et al. (1993) Neurotoxicity of a prion protein fragment. *Nature* 362: 543–546

13 Forloni G, Del Bo R, Angeretti N et al. (1994) A neurotoxic prion protein fragment induces rat astroglial proliferation and hypertrophy. *Eur J Neurosci* 6: 1415–1422

14 Salmona M, Forloni G, Diomede L et al. (1997) A neurotoxic and gliotrophic fragment of the prion protein increases plasma membrane microviscosity. *Neurobiol Dis* 4: 47–57

15 Tagliavini F, Prelli F, Ghiso J et al. (1991) Amyloid protein of Gerstmann-Sträussler-Scheinker disease (Indiana kindred) is an 11 kd fragment of prion protein with an N-terminal glycine at codon 58. *EMBO J* 10: 513–519

16 Tagliavini F, Prelli F, Porro M et al. (1994) Amyloid fibrils in Gerstmann-Sträussler-Scheinker disease (Indiana and Swedish kindreds) express only PrP peptides encoded by the mutant allele. *Cell* 79: 695–703

17 Chen SG, Teplow DB, Parchi P et al. (1995) Truncated forms of the human prion protein in normal brain and in prion diseases. *J Biol Chem* 270: 19173–19180

18 Parchi P, Castellani R, Capellari S et al. (1996) Molecular basis of phenotypic variability in sporadic Creutzfeldt-Jakob disease. *Ann Neurol* 39: 767–778

19 Piccardo P, Seiler C, Dlouhy SR et al. (1996) Proteinase-k-resistant prion protein isoforms in Gerstmann-Sträussler-Scheinker disease (Indiana kindred). *J Neuropathol Exp Neurol* 55: 1157–1163

20 Brown DR, Herms J, Kretzschmar HA (1994) Mouse cortical cells lacking cellular PrP survive in culture with a neurotoxic PrP fragment. *Neuroreport* 5: 2057–2060

21 Brown DR, Schmidt B, Kretzschmar HA (1996) Role of microglia and host prion protein in neurotoxicity of a prion protein fragment. *Nature* 380: 345–347

22 Haik S, Peyrin JM, Lins L et al. (2000) Neurotoxicity of the putative transmembrane domain of the prion protein. *Neurobiol Dis* 7: 644–656

23 Chabry J, Ratsimanohatra C, Sponne I et al. (2003) *In vivo* and *in vitro* neurotoxicity of the human prion protein (PrP) fragment p118–135 independently of PrP expression. *J Neurosci* 23: 462–469

24 Tagliavini F, Forloni G, Colombo L et al. (2000) Tetracycline affects abnormal properties of synthetic PrP peptides and PrP^{Sc} *in vitro*. *J Mol Biol* 300: 1309–1322

25 Bonetto V, Massignan T, Chiesa R et al. (2002) Synthetic miniprion PrP106. *J Biol Chem* 277: 31327–31334

26 Salmona M, Morbin M, Massignan T et al. (2003) Structural properties of Gerstmann-Sträussler-Scheinker disease amyloid protein. *J Biol Chem* 278: 48146–48153

27 Salmona M, Malesani P, De Gioia L et al. (1999) Molecular determinants of the physicochemical properties of a critical prion protein region comprising residues 106–126. *Biochem J* 342 (Pt 1): 207–214

28 Kunz B, Sandmeier E, Christen P (1999) Neurotoxicity of prion peptide 106-126 not confirmed. *FEBS Lett* 458: 65–68

29 Brown DR (1999) Comment on: Neurotoxicity of prion peptide 106–126 not confirmed by Beat Kunz, Erika Sandmeier, Philipp Christen. *FEBS Lett* 460: 559–560

30 Forloni G, Salmona M, Bugiani O, Tagliavini F (2000) Comment on: Neurotoxicity of prion peptide 106–126 not confirmed by Beat Kunz, Erika Sandmeier, Philipp Christen. *FEBS Lett* 466: 205–206

31 Ball HL, Mascagni P (1992) Purification of synthetic peptides using reversible chromatographic probes based on the Fmoc molecule. *Int J Pept Protein Res* 40: 370–379

32 Sarin VK, Kent SB, Tam JP, Merrifield RB (1981) Quantitative monitoring of solid-phase peptide synthesis by the ninhydrin reaction. *Anal Biochem* 117: 147–157

33 Kyte J, Doolittle RF (1982) A simple method for displaying the hydropathic

character of a protein. *J Mol Biol* 157: 105–132

34 Deleage G, Roux B (1987) An algorithm for protein secondary structure prediction based on class prediction. *Protein Eng* 1: 289–294

35 Krogh A, Larsson B, von Heijne G, Sonnhammer EL (2001) Predicting transmembrane protein topology with a hidden Markov model: Application to complete genomes. *J Mol Biol* 305: 567–580

36 Sreerama N, Woody RW (2000) Estimation of protein secondary structure from circular dichroism spectra: Comparison of contin, selcon, and cdsstr methods with an expanded reference set. *Anal Biochem* 287: 252–260

37 Shahar A, de Vellis J, Vernadakis A, Haber B (eds) (1989) *A dissection and tissue culture manual of the nervous system*, Alan R. Liss, Inc., New York

38 Liu Y, Peterson DA, Kimura H, Schubert D (1997) Mechanism of cellular 3-(4,5-dimethylthiazol-2-yl)-2,5-diphenyltetrazolium bromide (MTT) reduction. *J Neurochem* 69: 581–593

39 Chiesa R, Angeretti N, Lucca E et al. (1996) Clusterin (Sgp-2) induction in rat astroglial cells exposed to prion protein fragment 106-126. *Eur J Neurosci* 8: 589–597

40 Hartley DM, Walsh DM, Ye CP et al. (1999) Protofibrillar intermediates of amyloid beta-protein induce acute electrophysiological changes and progressive neurotoxicity in cortical neurons. *J Neurosci* 19: 8876–8884

41 Walsh DM, Klyubin I, Fadeeva JV et al. (2002) Naturally secreted oligomers of amyloid beta protein potently inhibit hippocampal long-term potentiation *in vivo*. *Nature* 416: 535–539

42 Bucciantini M, Giannoni E, Chiti F et al. (2002) Inherent toxicity of aggregates implies a common mechanism for protein misfolding diseases. *Nature* 416: 507–511

43 Kayed R, Head E, Thompson JL et al. (2003) Common structure of soluble amyloid oligomers implies common mechanism of pathogenesis. *Science* 300: 486–489

44 Soto C, Castano EM, Kumar RA et al. (1995) Fibrillogenesis of synthetic amyloid-beta peptides is dependent on their initial secondary structure. *Neurosci Lett* 200: 105–158

45 Telling GC, Scott M, Mastrianni J et al. (1995) Prion propagation in mice expressing human and chimeric PrP transgenes implicates the interaction of cellular PrP with another protein. *Cell* 83: 79–90

46 Forloni G, Angeretti N, Malesani P et al. (1999) Influence of mutations associated with familial prion-related encephalopathies on biological activity of prion protein peptides. *Ann Neurol* 45: 489–494

47 Barret A, Tagliavini F, Forloni G et al. (2003) Evaluation of quinacrine treatment for prion diseases. *J Virol* 77: 8462–8469

48 Forloni G, Bugiani O, Tagliavini F, Salmona M (1996) Apoptosis-mediated neurotoxicity induced by beta-amyloid and PrP fragments. *Mol Chem Neuropathol* 28: 163–171

49 Chiesa R, Piccardo P, Quaglio E et al. (2003) Molecular distinction between pathogenic and infectious properties of the prion protein. *J Virol* 77: 7611–7622

50 Tagliavini F, McArthur RA, Canciani B et al. (1997) Effectiveness of anthracycline against experimental prion disease in Syrian hamsters. *Science* 276: 1119–1122

51 Forloni G, Iussich S, Awan T et al. (2002) Tetracyclines affect prion infectivity. *Proc Natl Acad Sci USA* 99: 10849–10854

52 Harris DA (1999) Cellular biology of prion diseases. *Clin Microbiol Rev* 12: 429–444

53 Hegde RS, Tremblay P, Groth D et al. (1999) Transmissible and genetic prion diseases share a common pathway of neurodegeneration. *Nature* 402: 822–826

54 Kourie JI (2001) Mechanisms of prion-induced modifications in membrane transport properties: Implications for

signal transduction and neurotoxicity. *Chem Biol Interact* 138: 1–26

55 Bahadi R, Farrelly PV, Kenna BL et al. (2003) Channels formed with a mutant prion protein PrP(82–146) homologous to a 7-kda fragment in diseased brain of GSS patients. *Am J Physiol Cell Physiol* 285: C862–872

56 Brown DR, Kretzschmar HA (1997) Microglia and prion disease: A review. *Histol Histopathol* 12: 883–892

57 Hope J, Shearman MS, Baxter HC et al. (1996) Cytotoxicity of prion protein peptide (PrP106-126) differs in mechanism from the cytotoxic activity of the Alzheimer's disease amyloid peptide, a-beta-25–35. *Neurodegeneration* 5: 1–11

58 Deli MA, Sakaguchi S, Nakaoke R et al. (2000) PrP fragment 106–126 is toxic to cerebral endothelial cells expressing PrP^C. *Neuroreport* 11: 3931–3936

59 O'Donovan CN, Tobin D, Cotter TG (2001) Prion protein fragment PrP-(106–126) induces apoptosis via mitochondrial disruption in human neuronal SH-SY5Y cells. *J Biol Chem* 276: 43516–43523

60 Hanan E, Goren O, Eshkenazy M, Solomon B (2001) Immunomodulation of the human prion peptide 106–126 aggregation. *Biochem Biophys Res Commun* 280: 115–120

61 Della-Bianca V, Rossi F, Armato U et al. (2001) Neurotrophin p75 receptor is involved in neuronal damage by prion peptide-(106–126). *J Biol Chem* 276: 38929–38933

62 Thellung S, Villa V, Corsaro A et al. (2002) P38 MAP kinase mediates the cell death induced by PrP106-126 in the SH-SY5Y neuroblastoma cells. *Neurobiol Dis* 9: 69–81

63 Jobling MF, Stewart LR, White AR et al. (1999) The hydrophobic core sequence modulates the neurotoxic and secondary structure properties of the prion peptide 106–126. *J Neurochem* 73: 1557–1565

64 Fioriti L, Quaglio E, Massignan T et al. (2004) The neurotoxicity of prion protein (PrP) peptide 106–126 is independent of the expression level of PrP an is not medicated by abnormal PrP species. *Mol Cell Neurosci*; in press

65 Hegde RS, Mastrianni JA, Scott MR et al. (1998) A transmembrane form of the prion protein in neurodegenerative disease. *Science* 279: 827–834

66 Ma J, Wollmann R, Lindquist S (2002) Neurotoxicity and neurodegeneration when PrP accumulates in the cytosol. *Science* 17: 17

67 Chiesa R, Harris DA (2001) Prion diseases: What is the neurotoxic molecule? *Neurobiol Dis* 8: 743–763

14 Cyclic Amplification of Prion Protein Misfolding

Joaquin Castilla, Paula Saá and Claudio Soto

Contents

1   Introduction

The objective of the methodology described in this Chapter is to mimic the *in vivo* replication of prions in the test tube in an accelerated manner, so that a high level of amplification of PrP^{res} can be obtained in few hours. For this purpose the concept of PMCA (protein misfolding cyclic amplification) has recently been developed [1].

Methods and Tools in Biosciences and Medicine
Techniques in Prion Research, ed. by S. Lehmann and J. Grassi
© 2004 Birkhäuser Verlag Basel/Switzerland

PMCA consists of cycles of accelerated prion replication [1]. Each cycle is composed of two phases (Fig. 1). During the first phase the sample containing minute amounts of PrP^{res} and a large excess of PrP^C are incubated to induce growing of PrP^{res} polymers. In the second phase the sample is sonicated in order to break down the polymers, multiplying the number of nuclei. Therefore, after each cycle the number of "seeds" is increased in an exponential fashion (Fig. 1) [1]. The cyclic nature of the system permits the use of as many cycles as required to reach the amplification state needed for the detection of PrP^{res} in a particular sample. PMCA is conceptually analogous to DNA amplification by polymerase chain reaction (PCR). In both systems a template grows at the expense of a substrate in a cyclic reaction, which combines phases of growing and multiplication of the template units.

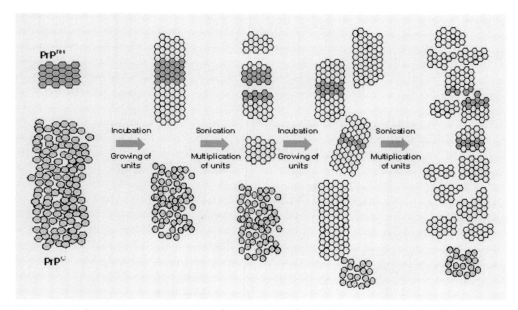

Figure 1   Schematic representation of the rational for PrP^{res} replication by PMCA.

In this Chapter we will describe in detail the technical aspects of PMCA and its application to different samples (blood, brain, spleen, etc.). In all cases, we will use a homologous healthy brain homogenate as source of PrP^C since other factors present in it might be important during prion replication [1]. The amount of PrP^C converted is evaluated by Western blotting after proteinase K (PK) treatment of samples following routine procedures.

2   Materials

Chemicals
- Deionized water (J.T. Baker, cat# 4201)
- NaCl (VWR, cat#: EM-SX0420–1)
- KCl (VWR, cat#: EM-PX1405–1)
- Na$_2$HPO$_4$ (VWR, cat#: 80503–412)
- KH$_2$PO$_4$ (VWR, cat#: 80503–252)
- SDS (VWR, cat#: 80503–358)
- Triton X-100 (VWR, cat#: 80503–490)
- PMSF (VWR, cat#: 80055–380)
- Complete Protease inhibitor (Roche, cat#: 1836145)
- Proteinase K (Roche, cat# 0745723)
- ETDA 0.5 M (pH 8.0) (Promega, cat # V4231)

Biologicals
- Transmissible spongiform encephalopathy (TSE) infectious material
- Normal brain

Solutions, reagents and buffers
- PBS 1X (Solution 1):

NaCl	8 g
KCl	0.2 g
Na$_2$HPO$_4$	1.44 g
KH$_2$PO$_4$	0.24 g

    Add H$_2$O
    Set up pH to 7.4
    Add H$_2$O up to 1,000 ml
- NaCl 5 M (Solution 2):

NaCl	58.44 g

    Add H$_2$O up to 200 ml
- Conversion buffer (Solution 3):

Phosphate buffered saline (PBS) 1X (Solution 1)	47.6 ml
NaCl 5 M (Solution 2)	1.5 ml
Triton X-100	0.5 ml
EDTA 0.5 M (pH 8.0)	0.4 ml
Complete Protease Inhibitor Cocktail	1 tablet (1×)
- SDS 4% (Solution 4):

Sodium dodecy sulphate (SDS)	4 g

    H$_2$O up to 100 ml
- Proteinase K 10 μg/μl (Solution 5):

Proteinase K (PK)	100 mg

    H$_2$0 up to 10 ml
    Note: Conserve this solution at –20 °C avoiding freezing-thawing.

- PMSF 5 mM (Solution 6):
  Phenylmethyl sulfonid fluoride (PMSF) 0.044 g
  Methanol up to 5 ml
- 10% infectious material homogenate (IMH):
  Infectious material   1 g
  Conversion buffer (Solution 3) up to 10 ml
  Note: The exact amount of infectious material used dependent upon the availability and type of material.
- 10% normal brain homogenate (NBH):
  Healthy brain from the same species   1 g
  Conversion buffer (Solution 3) up to 10 ml

Equipment
- Bandelin Sonoplus HD 2070 (Bandelin Electronic, Germany)
- Tip MS73 (Bandelin Electronic, Germany)
- Misonix 2020 sonicator (Misonix, USA)
- ULT 900 Upright Freezer –86 °C (ThermoForma, USA)
- pH meter Model 6171 (Jenco, USA)
- High-viscosity mixer (Homogenizer) Eurostar PWR BSC S1 (IKA, USA)
- Surgical equipment
- Thermomixer R (Eppendorf, USA)
- Pipettes (Rainin, USA)
- Centrifuge model 5414 (Eppendorf, USA)
- PCR tubes of 0.2 ml

3   Methods

It is important to highlight that as other emerging methodologies, the PMCA technique is continuously under improvement and optimization. This is because many variables (temperature, pH, substrate concentration, type and concentration of the detergents, power and duration of sonication, etc.) play an important role in the process. In this section we will provide a detailed description of the current methods used for PMCA, but the reader should be aware that the methodology is continuously changing and should be encouraged to introduce modifications enhancing the yield of PrP replication. Figure 2 schematizes the steps in the procedure.

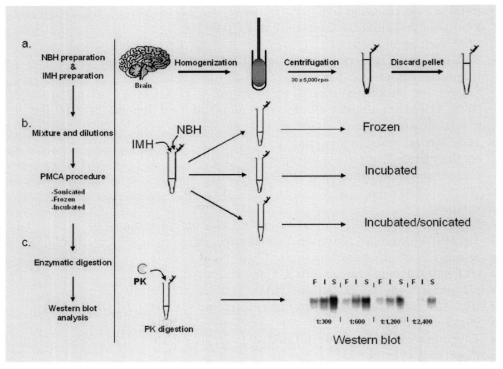

Figure 2   Scheme of the PMCA Procedure

3.1   Preparation of 10% (w/v) IMH inoculum

- Homogenate the infectious material in Solution 3 at 10% (w/v) at 4 °C using a high-viscosity mixer (see equipment).
- Centrifuge the homogenate at 2,000 g during 10 s and take the supernatant discarding the pellet. Store the clarified homogenate at –80 °C ready to use for PMCA.
  → See Figure 2a.

Note I: The most frequent IMH will be the infectious brain homogenate. Other tissue could be prepared in the same manner but may require other treatment depending upon the special characteristics of each tissue. Blood, and other tissues containing blood should be treated carefully since plasma or serum could inhibit the proteinase K treatment needed after amplification. It is recommended to avoid working with high amounts of blood in the samples used as inocula.

Note II: Although the inoculum does not require special storage procedure, it is recommended to store it frozen at –80 °C. It is possible to freeze-thaw it at least 20 times without significantly losing the efficiency.

3.2 Preparation of 10% (w/v) NBH substrate

- Prior to sacrifice, prepare animals with PBS containing 5 mM EDTA to remove as much as possible the blood.
- Sacrifice quickly the animal avoiding the use of anesthesia, as it may interfere with the amplification. The animal should be the same species than the prion sample to be amplified. It is recommended to use cervical dislocation or guillotine.
- Remove quickly the whole brain avoiding as much as possible the collection of blood. Wash the whole brain with cold PBS before the next step.
- Place the brain into solution 3 at 4 °C and homogenate the material at 10% (w/v) using a high-viscosity mixer (see equipment).
- Centrifuge the homogenate at 2,000 g during 10 s and take the supernatant, discarding the pellet. Store the clarified homogenate at −80 °C ready to use in PMCA. The weak centrifugation is important in order not to remove membrane components that are important for the conversion of PrP^C into PrP^{res}. → See Figure 2a.

Note I: The substrate should not be kept for a long time at 4 °C and freezing-thawing should be avoided. It is also highly recommended to store at −80 °C instead of −20 °C. Following the recent publication from Supattapone's group [19] it is recommended to work in RNAse free conditions.

Note II: The pH of the homogenate should range from 7 to 7.3.

3.3 Protein misfolding cycling amplification

PMCA can be done manually using single probe sonicator or automatically using a water-bath sonicator that can be programmed for automatic operation.

Manual PMCA
- Aliquots of 10% (w/v) IMH, or serial dilutions from it, are placed in 0.5 ml tubes at 4 °C and mixed with 80–100 µl of 10% (w/v) NBH from the homologous species. In each tube, the volume of NBH should be at least 50% (and ideally between 80–90%) of the total volume. Three equivalent tubes for each condition are prepared. One will be frozen immediately (frozen control), another will be incubated at 37 °C without sonication (incubated sample) and the third will be subjected to incubation/sonication cycles (PMCA samples) (see Fig. 2b).
- Incubate the tubes for 1 h at 37 °C with shaking set at 450 rpm using a Thermomixer R (see equipment).
- Every hour sonicate each tube with the following sequence: 10 pulses of 0.1 s at 0.9 s intervals.
- The process can be repeated as many times as needed to reach the desired level of amplification.

- After the last sonication the samples are ready to be digested. If digestion is not done immediately, it is recommended to store the amplified samples at −80 °C.

Note I: To analyze sensitivity of amplification and to calibrate PMCA technology, it is recommended to begin with a 1:100 dilution of 10% infectious brain homogenate and from there make 1:10 dilutions.

Note II: Although more studies have to be done, we recommend using the whole brain to prepare the homogenate.

Note III: The power of sonication and the incubation time between sonications in each cycle has to be adjusted for optimal amplification when using PrP^{res} obtained from different species/strains or from different tissues within the same species. For example, cycles of 1 h incubation followed by sonication pulses of around 40 watts are appropriate to amplify hamster 263 K prion, but 3 h incubation and a lower potency of 20 watts/pulse is optimal for the protein from mouse ME7. In all cases, the percentage of power is referred to Bandelin Sonoplus HD 2070 equipment and it is not clear how it extrapolates to the potencies of other instruments.

Note IV: For some samples where the concentration of PrP^{res} is very low (such as blood), it might be convenient to first carry out a larger incubation in order to allow sufficient time for small PrP^{res} oligomers to seed PrP conversion.

Note V: Although a cycle usually begin with the incubation and end with the sonication, for some samples where PrP^{res} is highly concentrated and aggregated (such as low dilutions of experimental scrapie brain homogenate), it is advisable to begin with a sonication instead of incubation.

Note VI: The process of sonication produces aerosols, leading to the lost of some material in each cycle. This protocol is planned to start with around 80 µl of solution and after 20 cycles, the volume should be around 60 µl, i.e., a loss of approximately 25% of the sample is expected. For safety conditions, it is recommended to perform the sonication in a closed container inside a BSL2 hood to avoid spreading the infectious material.

Note VII: To avoid cross-contamination, when changing the sample, the sonication tip should be washed by immersing it in water and performing pulses of sonication. After several uses, it is recommended to decontaminate the tip by wiping it with 2N NaOH.

Note VIII: It is highly recommended to do the whole amplification experiment without freezing-thawing the samples. If necessary, freezing should always be at −80 °C.

Automatic PMCA
- Aliquots of 10% IMH or dilutions from it are placed into 0.2 ml PCR tubes and mixed with 10% NBH. It is recommended to use volumes between 80–100 µl and in each condition NBH should represent at least 50% of the sample.
- Incubate the tubes for 30 min at 37 °C in the water bath of the automatic sonicator (see equipment).

- Sonicate the tubes for 1 pulse of 40 s.
- The cyclical process is repeated as many times as needed to reach a desirable level of amplification.

Note I: It is very important to use thin wall tubes in order to obtain the most effective penetration of ultrasound waves.

Note II: The power of sonication for Hamster 263 K should be set to the 4–8 potency of this sonicator. For other species/strains sonication power should be optimized experimentally.

Note III: The reservoir has to be filled with 140 ml of water, which decreases at a rate of around 1 mm/4 h at 37 °C. It is recommended to use sterile deionized water. It is also recommended to incubate the tubes without touching the sonication plate.

Note IV: The highest amplification is obtained using 10% of substrate but it is also possible to use 2.5–5%, although with a slight decrease in the amplification yield.

Note V: In the conditions described in this protocol the PrP^C is known to be stable for at least 100 h at 37 °C, but other essential yet unknown factors could be degraded before.

3.4 PrP^{res} detection

- To each tube add 1 μl of solution 4.
- Incubate the samples with PK (solution 5) during 1 h at 45 °C using the standard concentration depending of the prion protein used (i. e., 50 μg/ml for hamster, 20 μg/ml for cattle or 40–50 μg/ml for variant Creutzfeldt-Jakob disease (vCJD)).
- Stop the PK digestion with 1 μl of solution 6.
- Add SDS-PAGE loading buffer to prepare the samples for Western blot analysis.
  → See Figure 2c.

4   Troubleshooting

In this section we will provide guidelines to attempt resolving some of the most common problems that we and others have encountered when using the PMCA technique.

4.1 Problem 1: No amplification

Following are some of the most important aspects to be considered when no amplification is obtained:

NBH: A good substrate is the key to get the best amplification. The tissue has to be fresh or frozen without fixation and taken with the shortest possible delay after death. After homogenization, large pieces of tissue and entire cells are removed by a low-speed centrifugation. The low speed is important in order not to remove/destabilize some membrane components which seem essential for conversion. The solution should be turbid with clear pieces of membrane in it. If the homogenate is transparent the efficiency of conversion will not be good. In addition, the NBH should not be kept for a long time at 4 °C. Freezing-thawing should also be avoided (storing at –80 °C instead of –20 °C is highly recommended). The level of amplification could dramatically decrease using a NBH prepared under conditions other than those suggested. In view of the recent paper from Supattapone's group [2], working in RNAse free conditions is recommended.

IMH: The amount of PrP[res] in the sample to be amplified is the most important issue to be considered since this quantity will determine the number of cycles needed to detect PrP[res] after PMCA.

Sonication conditions: An optimal power of sonication is crucial to cut down the PrP[res] polymers without affecting their capability of acting as nuclei to further convert PrP[C]. A too-weak sonication might not be efficient in cutting the aggregates and a too-strong sonication might physically damage the protein polymers minimizing their capability to grow further. In our experience conventional water bath sonicators do not have enough potency to produce optimal PMCA amplification. Unfortunately, there is no clear way to standardize sonication power among different machines and thus the best sonication conditions should be experimentally identified in each apparatus. In order to find out if sonication power is the problem in the lack of amplification, it is recommended to run a Western blot with equivalent samples incubated (without sonication) for the same time. At low dilutions of PrP[res] containing brain homogenate, incubation should give a modest but clear amplification of the signal (usually less than 10-fold). If no increase in the signal is seen with incubation alone, it is likely that the problem is something else besides sonication power. To begin PMCA we recommend first standardizing the conditions to reach at least >3-fold amplification by incubation of the samples for 10 h.

It is also important to be fully aware that the ultrasound strength needed for amplification of PrP[res] samples from different species/strains and sources of PrP[res] can be different, and hence low or even no amplification might be seen for new samples under conditions that work very well for others. These findings are probably related to the specific conformation/aggregation state of each strain of prions, which has been proposed to explain the differences in clinical, pathological and biochemical features of distinct strains.

Conversion buffer (Solution 3): Although different buffers might work in PMCA methodology, exhaustive studies have demonstrated that even small changes in the composition of the buffer dramatically affect the amplification process. For example, pH's of around 7–7.3 are necessary to obtain the best results using this technique.

4.2 Problem 2: Low amplification

In addition to the technical considerations described above, other methodological details have to be taken into account when a low amplification is obtained.

Number of cycles: There is a clear relationship between the level of amplification and the number of cycles used. Obviously, the methodology used for the amplification, whether manual or automatic, determines to a large extent the feasible number of cycles. When many cycles are carried out another limitation appears: the stability of PrP^C and other co-factors at 37 °C in NBH. PrP^C and co-factors might be lost because of degradation, denaturation or aggregation and it is likely that during long incubations these three processes may be operating simultaneously. Our data indicate that at least 90% of PrP^C remains non-degraded during 100 h at 37 °C using a standard NBH when the indicated amount of protease inhibitors are used. However, since the identity of other co-factors is unknown, it is not possible to directly evaluate the loss of these important factors. One approach to minimize the loss of PrP^C and/or co-factors is to add aliquots of NBH every certain number of cycles. Although this improves PrP^{res} formation when many cycles are performed, it complicates the interpretation of the results, especially when quantitative studies are being performed. Another complication is freezing-thawing during PMCA, which is especially complicated when many cycles are performed using manual sonicator. It is clear that freezing-thawing between cycles of amplification yields lower levels of amplified PrP^{res}.

pH: The best amplification is obtained using a conversion buffer between pH 7–7.3. The use of conversion buffers with pHs out of this range could reduce the level of amplification.

Temperature: The best amplification is obtained when the range of temperature is between room temperature and 37 °C. Incubations at temperatures higher than 37 °C during PMCA could reduce the level of amplification.

Type of detergent: Although other detergents such as 1% of Nonidet-P40 also work for PMCA, the use of 1% of Triton X-100 is highly recommended. Small concentrations of SDS also work for PMCA but often are not needed and its percentage should be adapted depending on the type of the PrP^{res} species being amplified.

Substrate concentration: We have obtained the best PMCA amplifications of PrP^{res} using substrate concentrations ranging from 2.5% to 10%. Lower levels of substrate yield low levels of amplification and percentages higher than 10%

usually present problems with PK digestion. The recommended concentration in standard studies is 10% independently of the species used.

Volumes: Although the volume should not be an important issue to be considered, the use of high volumes of sample could decrease the effective power of sonication in the homogenate resulting in lower levels of amplification. All the data shown in this Chapter corresponds to volumes between 80–100 μl. The use of smaller volumes favor the loss of sample reducing the yield of the amplification.

Length of the pulses: This is a difficult issue since more information is necessary to understand the internal process of sonication. According to empirical results at least 10 pulses of 0.1 s each are recommended in manual sonication and at least 40 s of sonication in automatic sonication.

Tune of sonication: Although this issue is not directly related with the PMCA methodology, it is recommended to check the tune of the sonicator since a bad tune could reduce the effective power of the sonicator resulting in a decrease of the yield. Also, the efficiency of sonication goes down when the sonicator tip has been used for a long time.

4.3   Problem 3: High variability and lack of linearity in the results

Sonication power: Occasionally, the high power of sonication may splash the sample on the lid of the tube reducing the effect of sonication. This effect may also occur with low sample volumes, low substrate concentration and low levels of bath water in the automatic sonicator. Important differences among tubes containing identical samples may also appear if bubbles are made during sonication.

Tune of sonication in all tubes of plaque: Checking that all the 96 tubes in the automatic sonicator are receiving the same sonication power is very important to minimize variability. Good tuning is essential to reduce the variability in the results and to obtain similar yielding in all the sonicated tubes. When tuning is properly controlled no significant differences are seen among tubes placed at distinct positions of the plaque.

Formation of large aggregates: When high amounts of PrP^{res} are produced, it is possible that large and stable aggregates may be formed during PMCA. These aggregates might not be accessible to PK, resulting in an incomplete digestion, and they might be resistant to boiling in SDS and hence migrate in electrophoresis as large molecular weight species that do not blot efficiently. To minimize this problem, we recommend a last sonication of 10 pulses at 100% of power in the manual sonicator previous to PK digestion or adding a higher amount of SDS (0.25%) before protease treatment.

4.4   Problem 4: Inappropriate PK digestion

Large aggregates: See recommendations above.

Blood: The blood contains protease inhibitors that could interfere with the PK digestion. Avoid working with high amounts of blood during the extraction of the brain used as NBH.

PK condition: Since different species present different PrPres characteristics, the optimal PK treatment condition should be previously checked. Large incubation times will occasionally require higher concentrations of PK as a consequence of the formation of big aggregates. Increasing the concentration of SDS in the buffer used for the PK treatment could be necessary. In addition, working with temperatures between 37–64 °C and shaking at 350–450 rpm are also recommended.

5   Applications of PMCA

There are several different possible applications for the PMCA technology as a sensitive *in vitro* model to help understand the underlying biology of prions, to identify other factors that may be responsible for prion protein conversion, to discover novel drug targets for TSEs and to develop a highly-sensitive TSE diagnosis.

PMCA provides a great opportunity to evaluate the infectious properties of PrPres generated *in vitro*, because, after amplification using the optimal conditions the total amount of resultant protease-resistant protein is mainly composed of newly produced PrPres (>99%). The latter is essential to distinguish the infectivity coming from the inoculum from newly generated infectivity.

Another widely debated issue in the TSE field is the molecular mechanism of species barrier and prion strains [3, 4]. As a consequence of the transmission of BSE to humans, a great concern has arisen regarding interspecies infectivity and the tissues having a quantity of prions high enough to transmit the disease [5, 6]. The molecular aspects that underly the species barrier phenomenon are still not understood. It has been shown that the sequence identity between the infectious PrPSc and the host prion protein plays a crucial role in determining species barrier [7]. It is clear that few amino acid differences between both proteins can modify dramatically the incubation time and the course of the disease [8, 9]. So far, the investigation of the species barrier and the tissues carrying infectivity has been done using the biological assay of prion propagation in animals [4, 5]. However, these studies are time consuming, because it is necessary to wait for several months or even years until the animals develop the clinical signs. In addition, the assessment of the species barrier for transmission of prions to humans is compromised by the use of animal models. PMCA can provide a complement to the *in vivo* studies of the phenomenon of species barrier by combining PrP[res] and PrP[C] from different sources in distinct quantities and evaluate quantitatively the efficiency of the conversion.

PMCA can also contribute to a better understanding of the mechanism of prion conversion and the identification of additional factors involved. Based on data obtained with transgenic animals, additional brain factors present in the host have been proposed to be essential for prion propagation [10, 11]. We reported previously that using highly purified prion proteins (PrP[res] and PrP[C]) the conversion procedure does not occur under our experimental conditions [12]. However, the activity is recovered when the bulk of cellular proteins are reincorporated into the sample [12]. This finding provides direct evidence that other factors present in the brain are essential to catalyze prion propagation. PMCA represents an ideal biochemical tool to identify these conversion factors. The identification and characterization of these factors will not only increase our knowledge of the molecular mechanism of prion conversion, but may also help to understand similar protein misfolding processes associated with other neurodegenerative diseases, such as Alzheimer's and Parkinsons disease [13].

An obvious application of PMCA is the TSE diagnosis. As stated before, the biggest problem facing a biochemical test to detect PrP[res] pre-symptomatically in tissues other than brain is the very low amount of PrP[res] existing in them. Most of the efforts to develop a diagnostic system for prion diseases have been focused on the increase of sensitivity of the current detection methods. PMCA offers the opportunity to enhance existing methods by amplifying the amount of PrP[res] in the sample. Combining the strategy of reproducing prions *in vitro* with any of the high-sensitive detection methods, the early diagnosis of TSE may be achieved. The aim would be not only to detect prions in the brain in early pre-symptomatic cases, but also to generate a test to diagnose living animals and humans. For this purpose a tissue other than brain is required and in order to have an easier non-invasive method, detection of prions in body fluids such as urine or blood are the best option. A blood test for CJD can have many applications, including screening of blood banks, identification of populations at risk, reduction of iatrogenic transmission of CJD and early diagnosis of the disease [14, 15].

PMCA can also have applications in the development of novel therapies for TSE. One of the best targets for TSE therapy is the inhibition and reversal of PrP[C] to PrP[res] conversion [16, 17]. In drug development it is crucial to have a relevant and robust *in vitro* assay to screen compounds for activity before testing them in more time consuming and expensive *in vivo* assays. PMCA represents a convenient biochemical tool to identify and evaluate the activity of drug candidates for TSE treatment, because it mimics *in vitro* the central pathogenic process of the disease. Also the simplicity of the method and the relatively rapid outcome are important features for this type of study. Moreover, the fact that PMCA can be applied to prion conversion in different species provides the opportunity to validate the use in humans of drugs that have been evaluated in experimental animal models of the disease.

6   Remarks and conclusions

PMCA marks the first time in which the (mis)folding and biochemical properties of a protein has been amplified cyclically in a similar manner as DNA amplification by PCR. It is also the first time in which prions have been replicated in the test tube with high efficiency. The potential applications of this technology are many and include the understanding of the molecular mechanism of prion replication, the nature of the infectious agent and the development of novel diagnosis and therapies for these insidious diseases.

Without doubt the prion phenomenon of transmission of biological information by replication of protein misfolding represents one of the most interesting new findings in biology. The critical issue to understand in the near future is whether the prion phenomenon is exclusively associated with the prion protein or whether it represents a more general process in biology [17]. The finding of proteins with a prion-like behavior in yeast and other fungi have provided a step forward in answering this question [18, 19]. We cannot predict precisely how common the prion phenomenon is in nature, but being able to reproduce it *in vitro* with high efficiency could give us a powerful tool to explore the amplification of protein (mis)folding using proteins not yet known to have a prion behavior.

Besides TSE, there are several other diseases involving changes in the conformation of a natural protein to an altered structure with toxic properties capable of inducing tissue damage and organ dysfunction [13, 20]. This group of diseases, called protein conformational disorders, includes several forms of neurodegenerative diseases such as Alzheimer's, Parkinsons and Huntington's disease as well as a group of more than 15 distinct disorders involving amyloid deposition in diverse organs. In a similar way to PrP^{res} in TSE, the protein conformational changes associated with the pathogenesis of these diseases results in the formation of abnormal proteins rich in β-sheet structure, partially resistant to proteolysis and with a high tendency to aggregate [13]. The process of misfolding and aggregation also follows a seeding-nucleation mechanism and hence the principles of PMCA might be applied to amplify the abnormal folding of these proteins as well. Therefore, PMCA may have a broader application for research and diagnosis of diseases where misfolding and aggregation of a protein is a hallmark event.

Acknowledgements

This work is partially supported by a European Community grant (TSELAB) and by NIH grant R03 AG024642.

Further reading

Saborio GP, Permanne B, Soto C (2001) Cyclic amplification of protein misfolding: A novel approach for sensitive detection of pathological prion protein. *Nature* 411: 810–813

Soto C, Saborio GP, Anderes L (2002) Cyclic amplification of Protein Misfolding: Applications in prion research and beyond. *Trends Neurosci* 25: 390–394

Telling G (2001) Protein-based PCR for prion diseases? *Nat Med* 7: 778–779

Szallasi Z (2001) 'Scrapie-ing' together a signal from prions. *Trends Pharmacol Sci* 22: 447

Soto C (2002) Altering prion replication for therapy and diagnosis of Transmissible Spongiform Encephalopathies. *Biochem Soc Transactions* 30: 569–574

Soto C (2003) Unfolding the role of Protein Misfolding in Neurodegenerative Diseases. *Nature Rev Neurosci* 4: 49–60

Soto C, Saborio GP (2001) Prions: Disease-propagation and disease-therapy by conformational transmission. *Trends Mol Med* 7: 109–114

Soto C (2004) Diagnosing prion diseases: needs, challenges and hopes. *Nature Rev Microbiol* 2: 809–819

Soto C, Castilla J (2004) The controversial protein-only hypothesis of prion propagation. *Nature med* 10: 563–567

References

1   Saborio GP, Permanne B, Soto C (2001) Sensitive detection of pathological prion protein by cyclic amplification of protein misfolding. *Nature* 411: 810–813

2   Deleault NR, Lucassen RW, Supattapone S (2003) RNA molecules stimulate prion protein conversion. *Nature* 425: 717–720

3   Kascsak RJ, Rubenstein R, Carp RI (1991) Evidence for biological and structural diversity among scrapie strains. *Curr Topics Microbiol Immunol* 172: 139–150

4   Clarke AR, Jackson GS, Collinge J (2001) The molecular biology of prion propagation. *Philosophical Transactions of the Royal Society of London – Series B: Biological Sciences*: 356: 185–195

5   Wadsworth JD, Joiner S, Hill AF et al. (2001) Tissue distribution of protease resistant prion protein in variant Creutz-feldt-Jakob disease using a highly sensitive immunoblotting assay. *Lancet* 358: 171–180

6   Hill AF, Joiner S, Linehan J et al. (2000) Species-barrier-independent prion replication in apparently resistant species. *Proc Natl Acad Sci USA* 97: 10248–10253

7   Telling GC, Parchi P, DeArmond SJ et al. (1996) Evidence for the conformation of the pathologic isoform of the prion protein enciphering and propagating prion diversity. *Science* 274: 2079–2082

8   DeArmond SJ, Prusiner SB (1996) Transgenetics and neuropathology of prion diseases. *Curr Top Microbiol Immunol* 207: 125–146

9   Asante EA, Collinge J (2001) Transgenic studies of the influence of the PrP structure on TSE diseases. *Adv Prot Chem* 57: 273–311

10 Telling GC, Scott M, Mastrianni J et al. (1995) Prion propagation in mice expressing human and chimeric PrP transgenes implicates the interaction of cellular PrP with another protein. *Cell* 83: 79–90

11 Kaneko K, Zulianello L, Scott M et al. (1997) Evidence for protein X binding to a discontinuous epitope on the cellular prion protein during scrapie prionápropagation. *Proc Natl Acad Sci USA* 94: 10069–10074

12 Saborio GP, Soto C, Kascsak RJ et al. (1999) Cell-lysate conversion of prion protein into its protease-resistant isoform suggests the participation of a cellular chaperone. *Biochem Biophys Res Commun* 258: 470–475

13 Soto C (2001) Protein misfolding and disease; protein refolding and therapy. *FEBS Lett* 498: 204–207

14 Schiermeier Q (2001) Testing times for BSE. *Nature* 409: 658–659

15 Anonymous (2001) Scientists race to develop a blood test for vCJD. *Nat Med* 7: 261

16 Head MW, Ironside JW (2000) Inhibition of prion protein conversion: a therapeutic tool?. *Trends Microbiol* 6: 6–8

17 Soto C, Saborio GP (2001) Prions: Disease propagation and disease therapy by conformational transmission. *Trends Mol Med* 7: 109–114

18 Wickner RB, Edskes HK, Maddelein ML et al. (1999) Prions of yeast and fungi Proteins as genetic material. *J Biol Chem* 274: 555–558

19 Lindquist S (1997) Mad cows meet psichotic yeast: the expansion of the prion hypothesis. *Cell* 89: 495–498

20 Carrell RW, Lomas DA (1997) Conformational disease. *Lancet* 350: 134–138

Guide to Solutions

Guide to Protocols

Troubleshooting Guide

Index

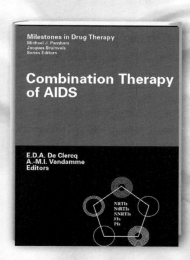

De Clercq, E.D.A., Katholieke Universiteit Leuven, Belgium / Vandamme, A.-M., Katholieke Universiteit Leuven, Belgium (Eds.)

Combination Therapy of AIDS

2004. 243 pages. Hardcover
ISBN 3-7643-6600-1
MDT - Milestones in Drug Therapy

Combination antiretroviral therapy of AIDS is a subject that has recently become a discipline of its own. For the first time, all facets are considered in a single volume. Combination drug treatment of AIDS is examined from different perspectives, including the history and basis, economic and theoretical considerations. Virological, immunological, clinical and practical reasons for therapy failures are given as well as strategies for preventing or overcoming therapy failures.

This work makes an important contribution to the current literature on clinical therapeutics. It is of interest for clinicians, researchers, public health workers, specialists and non-specialists seeking to optimize the treatment of AIDS.

From the contents:
● Anti-HIV agents to be used in drug combination regimens
● A perspective of the history of HAART
● The basic principles for combination therapy
● Comparison of the efficacy of HAART: single, dual or triple-class antiretroviral therapy
● Pharmacokinetics and pharmacodynamics of HAART
● Primary HIV infection: from diagnosis to treatment
● Salvage therapy
● Structured therapy interruptions (STIs): Lessons from a therapeutic strategy

For orders originating from all over the world except USA and Canada:
Birkhäuser Verlag AG
c/o Springer GmbH & Co
Haberstrasse 7
D-69126 Heidelberg
Fax: +49 / 6221 / 345 4 229
e-mail: birkhauser@springer.de
http://www.birkhauser.ch

For orders originating in the USA and Canada:
Birkhäuser
333 Meadowland Parkway
USA-Secaucus
NJ 07094-2491
Fax: +1 201 348 4505
e-mail: orders@birkhauser.com